Springer Tracts in Mechanical Engineering

Springer Tracts in Mechanical Engineering (STME) publishes the latest developments in Mechanical Engineering - quickly, informally and with high quality. The intent is to cover all the main branches of mechanical engineering, both theoretical and applied, including:

- Engineering Design
- Machinery and Machine Elements
- Mechanical Structures and Stress Analysis
- Automotive Engineering
- Engine Technology
- Aerospace Technology and Astronautics
- Nanotechnology and Microengineering
- Control, Robotics, Mechatronics
- MEMS
- Theoretical and Applied Mechanics
- Dynamical Systems, Control
- Fluids Mechanics
- Engineering Thermodynamics, Heat and Mass Transfer
- Manufacturing
- Precision Engineering, Instrumentation, Measurement
- Materials Engineering
- Tribology and Surface Technology

Within the scope of the series are monographs, professional books or graduate textbooks, edited volumes as well as outstanding PhD theses and books purposely devoted to support education in mechanical engineering at graduate and post-graduate levels.

Indexed by SCOPUS, zbMATH, SCImago.

Please check our Lecture Notes in Mechanical Engineering at https://link.springer.com/bookseries/11236 if you are interested in conference proceedings.

To submit a proposal or for further inquiries, please contact the Springer Editor **in your region**:

Ms. Ella Zhang (China)
Email: ella.zhang@springernature.com
Swati Meherishi (India)
Email: swati.meherishi@springer.com
Dr. Leontina Di Cecco (All other countries)
Email: leontina.dicecco@springer.com

All books published in the series are submitted for consideration in Web of Science.

Vitalii Ivanov · Artem Evtuhov · Ivan Dehtiarov ·
Justyna Trojanowska

Fundamentals of Manufacturing Engineering Using Digital Visualization

Springer

Vitalii Ivanov
Department of Manufacturing Engineering, Machines and Tools
Sumy State University
Sumy, Ukraine

Artem Evtuhov
Department of Manufacturing Engineering, Machines and Tools
Sumy State University
Sumy, Ukraine

Ivan Dehtiarov
Department of Manufacturing Engineering, Machines and Tools
Sumy State University
Sumy, Ukraine

Justyna Trojanowska
Faculty of Mechanical Engineering
Poznan University of Technology
Poznan, Poland

ISSN 2195-9862 ISSN 2195-9870 (electronic)
Springer Tracts in Mechanical Engineering
ISBN 978-3-031-74359-7 ISBN 978-3-031-74360-3 (eBook)
https://doi.org/10.1007/978-3-031-74360-3

This work was supported by Polish National Agency for Academic Exchange.

This Springer imprint is published by the registered company Springer Nature Switzerland AG
The registered company address is: Gewerbestrasse 11, 6330 Cham, Switzerland

Preface

Manufacturing engineering is at the heart of industrial innovation, influencing everything from creating simple consumer products to developing complex machinery. In the rapidly evolving field of manufacturing engineering, fundamentals equip professionals with the knowledge and skills to design, analyze, and improve manufacturing processes. By integrating digital visualization techniques into education, we empower engineers to understand better and optimize these processes, enhancing efficiency and productivity.

Chapter 1 introduces the role of manufacturing engineering, tracing its historical development and projecting future trends likely to shape the industry. This foundational chapter sets the stage for a deeper exploration of manufacturing principles.

Chapter 2 offers an overview of the manufacturing landscape, examining the various types of products, manufacturing processes, and production lines that define modern industry. The formation of technological assembly schemes is explained in detail. Moreover, the critical steps for preparing manufacturing and comprehensive classification of technological processes are described.

Chapter 3 discusses locating, a crucial aspect of manufacturing precision. This chapter covers the fundamental principles of locating, the classification of locating methods, and the design, process, and measuring datums that ensure accuracy in production. Real-world case studies are presented to illustrate the practical application of these concepts.

Chapter 4 focuses on product quality, with a particular emphasis on accuracy indicators, machining accuracy, and surface roughness. The chapter also explores the impact of surface quality on the exploitation properties of manufactured products, underscoring the importance of accuracy in every stage of the manufacturing process.

Chapter 5 describes the technological processes in machining, outlining the fundamental principles and strategies for determining workpiece machining routes. This chapter also discusses selecting appropriate machine tools and fixtures, providing a strategic approach to optimizing machining processes.

The final chapter, Chap. 6, presents a detailed examination of various machining methods, including turning, drilling, milling, grinding, and gear machining. Each method is discussed in depth, emphasizing the practical application of these techniques in the manufacturing industry.

Engineering education is essential for fostering innovation and solving complex problems in today's rapidly advancing technological landscape. In manufacturing engineering, it is crucial to present fundamental concepts in a way that is both accessible and engaging. This book integrates QR codes that can be scanned using a specially developed mobile application. These QR codes link to interactive digital content that simplifies complex topics, making them easier to understand and apply. By combining traditional learning with modern digital tools, this book provides a comprehensive and intuitive approach to mastering manufacturing engineering fundamentals.

A new mobile application, ManuVis, is specifically designed to enhance the engineering education experience for students and professionals. Available for Android and iOS devices, this app aims to make learning more accessible, interactive, and practical. The app provides a digital tool for visualizing key concepts in manufacturing engineering, offering an interactive way to explore product structure, assembly processes, technological assembly schemes, and locating methods. ManuVis makes the learning experience seamless and enjoyable with detailed scenarios and intuitive designs. The app can be easily downloaded using the QR codes below, which provide direct links to the Google Play and AppStore markets.

ManuVis
Manufacturing Visualization
GET IT ON Google Play

ManuVis
Manufacturing Visualization
Available on the App Store

This mobile application represents a significant step forward in engineering education, making it more accessible, flexible, and engaging.

This book aims to bridge the gap between traditional engineering concepts and modern digital tools, providing a comprehensive guide for students, professionals, and curious learners interested in mastering manufacturing engineering.

Sumy, Ukraine	Vitalii Ivanov
Sumy, Ukraine	Artem Evtuhov
Sumy, Ukraine	Ivan Dehtiarov
Poznan, Poland	Justyna Trojanowska
August, 2024	

Acknowledgments This book was prepared within the project "Digitalization in mechanical engineering" (agreement no. BNI-UE-2023-2) realized by Poznan University of Technology (Poland) and Sumy State University (Ukraine) and funded by the Polish National Agency for Academic Exchange (NAWA). It was also partially supported by International Association for Technological Development and Innovations and International Innovation Foundation.

Contents

About the Authors

Prof. Vitalii Ivanov DSc., Ph.D., ING.PAED.IGIP, is Head of the Department of Manufacturing Engineering, Machines and Tools, Director of the Research and Educational Center for Industrial Engineering at Sumy State University, and Vice Dean for International Affairs at the Faculty of Technical Systems and Energy Efficient Technologies of Sumy State University. He was awarded Scholarship of the Cabinet of Ministers of Ukraine for Young Scientists (2011 and 2014), Fellowship for a Research Internship from the Ministry of Education and Science, Youth and Sports of Ukraine (2011 and 2012), National Scholarship of the Slovak Republic (2015, 2017, and 2022), JICA Knowledge Co-Creation Fellowship (2016), the Ulam NAWA Program (2022), and State Prize of the Supreme Council of Ukraine for Young Scientists (2022). Prof. Ivanov carried out research, completed internships, and gave lectures in Germany, the USA, Slovak Republic, Poland, Great Britain, Italy, Japan, Czech Republic, Romania, China, etc. Furthermore, he participated in research projects funded by the Ministry of Education and Science of Ukraine, State Fund for Fundamental Research, and Ukrainian and German industrial companies. He also has experience coordinating and co-coordinating research projects funded by TEMPUS, Erasmus+, Czech Development Agency, DAAD, Gene Haas Foundation & Haas Automation, Japan International Cooperation Agency, and NAWA. Since 2019, Prof. Vitalii Ivanov has been Member of the Advisory Board for Research Evaluation at the University of West Bohemia (Czech Republic). Since 2020, he has been Member of the Experts' Team of the Invention Support Fund of the Ministry of Economic Development and Trade of Ukraine and Expert of the Ministry of Education and Science of Ukraine on the State Certification of Scientific Institutions. In 2022, he became Member of the Expert Council of the Ministry of Education and Science of Ukraine on certification of scientific personnel in Manufacturing Engineering. In 2021, Prof. Ivanov became International Member of the Academic Committee of the Doctoral Program "Innovative Engineering Technologies for Industrial Sustainability—IETIS" at the University of Salerno (Italy). In 2022, Prof. Ivanov became Member of the Specialized Councils for obtaining a Doctor of Science Degree in 05.02.08—Machine Building Technology,

05.03.01—Machining Processes, Machines, and Tools at National Technical University "Kharkiv Polytechnic Institute" (Ukraine) and Odesa Polytechnic National University (Ukraine). Since 2024, he has become Expert in carrying out scientific evaluation in manufacturing and mechanical engineering (Ministry of Education and Science of Ukraine).

Dr. Artem Evtuhov Ph.D., is Vice Dean for the educational and organizational work of the Faculty of Technical Systems and Energy Efficient Technologies and Associate Professor of the Department of Manufacturing Engineering, Machines and Tools at Sumy State University (Sumy, Ukraine). Artem Evtuhov is Member of the International Association for Technological Development and Innovations (Ukraine) and the International Association of Engineers (Hong Kong). He is Author of more than 50 scientific papers, patents, and didactic materials. His research activity deals with the development of manufacturing systems and technological processes.

Dr. Ivan Dehtiarov Ph.D., is Associate Professor of the Department of Manufacturing Engineering, Machines and Tools, Sumy State University (Sumy, Ukraine). Ivan Dehtiarov is Member of the International Association for Technological Development and Innovations (Ukraine) and the International Association of Engineers (Hong Kong). He is Author of more than 60 scientific papers, patents, and teaching materials. His research activities deal with the development of fixtures and technological equipment, as well as the programming of CNC machines.

Dr. Justyna Trojanowska Ph.D. Eng., is Doctor of Engineering in the scientific discipline of mechanical engineering, Assistant Professor at Poznan University of Technology (Poznan, Poland), Vice-Dean for cooperation with industry at the Faculty of Mechanical Engineering, and Editorial Board Member of the Lecture Notes in Mechanical Engineering (Springer). Her research activities concern production scheduling, production management, lean manufacturing, theory of constraints, and six sigma. She is also Member of the Polish Association for Production Management, the Polish Economic Society, the International Association for Technological Development and Innovations, and the Product Development and Management Association. Author of numerous publications in the area of mechanical engineering and production engineering, Dr. Trojanowska is participating in numerous international conferences and acting as Reviewer of scientific articles published in ISI master journals. She is also Member of the scientific committees of various international conferences. She is cooperating with foreign research centers and taking part in projects and scientific works conducted for production companies and aimed at improving production processes. She has also been one of the responsible persons for work package 5 "Training, Research & Development for Industry-Oriented Problems", aimed at reinforcing University-Business cooperation, and organized by EUNICE "European University for Customised Education".

Chapter 1
Introduction to Manufacturing Engineering

1.1 Role of Manufacturing Engineering

"Manufacturing" can be defined as transforming raw materials and components into finished products through various technological methods and operations. Manufacturing may include machining, chemical processing, assembly, and other processes that result in goods for use or sale.

The term "manufacturing" has its roots in Middle French and means "process of making". Originally, this word came from the Classic Latin "manu" and Middle French "facture". The earliest usage of the word manufacturing in the English language was recorded in the mid-sixteenth century to refer to making products by hand (Oxford English Dictionary 2010; Srivatsan et al. 2018). There are different interpretations of the term, summarized in Table 1.1.

Manufacturing engineering is critical for the modern economy and technological progress, as well as for ensuring sustainable development and global competitiveness (Fig. 1.1).

Manufacturing engineering is the basis for all other industries. It involves the production of tools, machinery, and equipment necessary to create a variety of goods, from consumer commodities to heavy machinery (Katina et al. 2023; Mishra et al. 2014). Manufacturing engineering brings new technologies and innovations to production processes, including automation, robotics, and the introduction of advanced manufacturing technologies that improve efficiency and product quality (Levinson 2017). Manufacturing engineering provides jobs and stimulates economic development. Engineering is also critical for exports, contributing to economic stability and growth. Mechanization and automation of production processes help to increase productivity. This reduces costs and increases production volumes, which is key to competitiveness in the global market (Relich et al. 2022). The machine-building industry has a multiplier effect on other sectors of the economy, including construction, transport, energy, and information technology (Novinkina et al. 2021).

V. Ivanov et al., *Fundamentals of Manufacturing Engineering Using Digital Visualization*, Springer Tracts in Mechanical Engineering,
https://doi.org/10.1007/978-3-031-74360-3_1

Table 1.1 Definition of manufacturing

Cambridge Business English Dictionary (2024)	The business of producing goods in large numbers, especially in factories
Merriam-Webster Dictionary (2003)	The process of making products, especially with machines in factories
Investopedia (Kenton 2024)	The processing of raw materials or parts into finished goods using tools, human labor, machinery, and chemical processing
Oxford Learner's Dictionaries (2024)	The business or industry of producing goods in large quantities in factories, etc
Collins Dictionary (2024)	The business of making things in factories
CIRP Encyclopedia of Production Engineering (2014)	The entirety of interrelated economic, technological, and organizational measures directly connected with the processing/machining of materials, i.e., all functions and activities directly contributing to the making of goods
CFI Education Inc. (Corporate Finance Institute 2024)	Processing of finished products from raw materials using various methods, human labor, and equipment according to a detailed plan in a cost-effective way
National Science Foundation (America's Investment in the Future 2000)	The process of converting dreams into objects that enrich lives
360 Degree Property and Engineering Services (Degree Property Engineering Services PVT LTD 2024)	The process of converting raw materials, components, or parts into finished goods that meet a customer's expectations or specifications. It involves a series of steps, including design, production, and quality control, to create products that can be distributed and sold
Components Engine s.r.l. (2024)	The production of products for use or sale using labor and machines, tools, chemical and biological processing, or formulation

Innovations in engineering often lead to progress in these industries, stimulating overall technological development (Muzylyov and Shramenko 2019).

Manufacturing plays a fundamental role in creating our modern world (Fig. 1.2). Expanding on this statement involves considering the pervasive influence of manufacturing across various sectors and its impact on economic, technological, and societal developments (World Manufacturing Report 2022). Trade in manufactured goods represented 71% of world merchandise exports (WTO: World Trade Report 2021). The manufacturing sector directly employs 13.6% of workers (International Labour Organization 2022). Manufacturing value added represents 17% of the global gross domestic product (World Bank 2022).

Based on recent statistics, the TOP-10 manufacturing countries are China, the USA, Germany, Japan, South Korea, and others mentioned in Fig. 1.3. Each

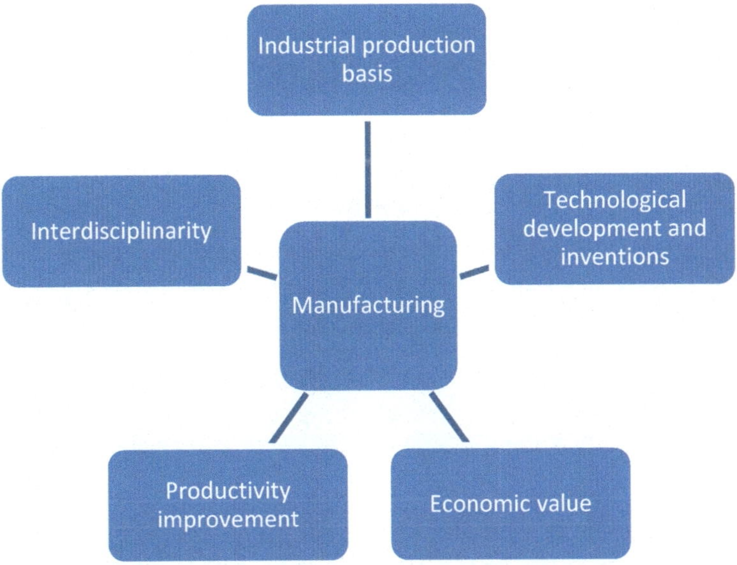

Fig. 1.1 Role of manufacturing

Fig. 1.2 Nothing happens without manufacturing

country has its specific focus. For instance, Japan is well-known for its automotive, electronics, robotics, and precision machinery.

Modern manufacturing faces many challenges in a dynamic and competitive global landscape. Several key factors are crucial for successfully overcoming the challenges of modern manufacturing:

- quality (ability to complete tasks according to assigned quality standards);
- speed (ability to complete tasks quickly and follow expectations);
- cost (ability to control operational costs and deliver within planned budget);
- flexibility (ability to react to new requests and requirements);

Fig. 1.3 TOP-10 major manufacturing countries worldwide

- dependability (ability to deliver tasks according to reliable approaches);
- innovation (ability to develop new and creative solutions).

These factors are interconnected and mutually reinforcing, allowing manufacturers to navigate complexities, remain competitive, and thrive in an ever-evolving industrial landscape. Embracing these principles contributes not only to short-term success but also to long-term sustainability and growth.

Manufacturing provides diverse skills and career opportunities. The sector encompasses various activities, from traditional production processes to advanced technologies, offering various career paths for individuals with different skill sets. The statistics (United Nations Economic Commission for Europe 2024) presented in Fig. 1.4. Prove the vital role of manufacturing.

The importance of manufacturing for society is profound, serving as a cornerstone for economic development, technological advancement, and employment opportunities.

Manufacturing produces goods that satisfy essential needs and desires, contributing significantly to a nation's GDP. It fosters innovation, propels technological progress, and plays a pivotal role in global trade. The sector creates jobs across various skill levels, driving social mobility and contributing to the overall well-being of communities.

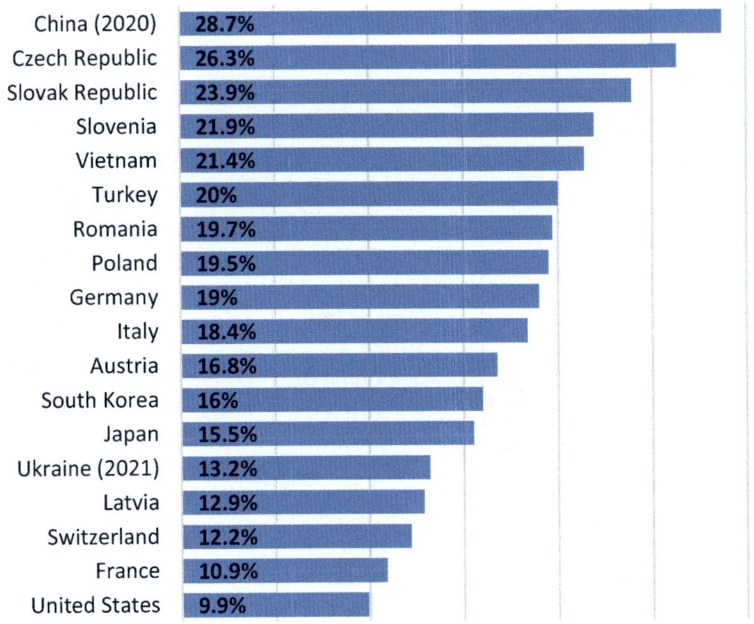

Fig. 1.4 Manufacturing employment as a proportion of total employment in 2022

Manufacturing enhances a country's competitiveness and resilience, ensuring a diversified and robust economy. As technology continues to reshape the manufacturing landscape, the sector remains a vital force, shaping the present and future prosperity of societies worldwide.

1.2 Historical Development of Manufacturing

The historical development of manufacturing engineering covers several stages, from primitive manual production to modern high-tech manufacturing processes. A brief overview of the key stages in the development of mechanical engineering (Fig. 1.5) is provided in this section.

The main features of "The Early Modern and Middle Ages Epoch" include using simple tools and mechanisms, developing handicraft production, and gradually accumulating knowledge and innovation that laid the foundations for future industrial revolutions. This period can be classified into:

- prehistory and antiquity: the first tools and mechanisms were made by hand using natural materials, including stone and bronze tools, and simple mechanisms, such as levers and wheels;

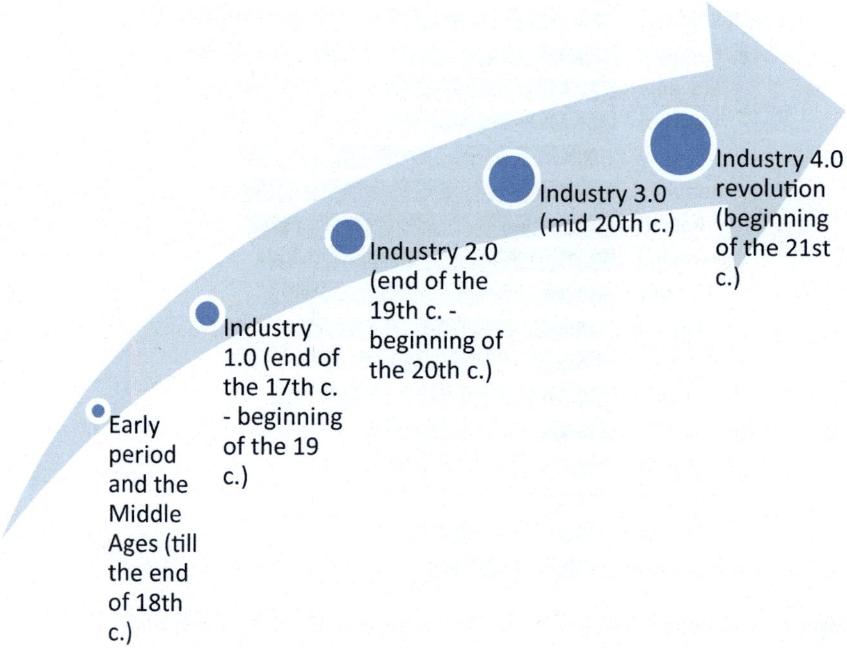

Fig. 1.5 Evolution of manufacturing engineering

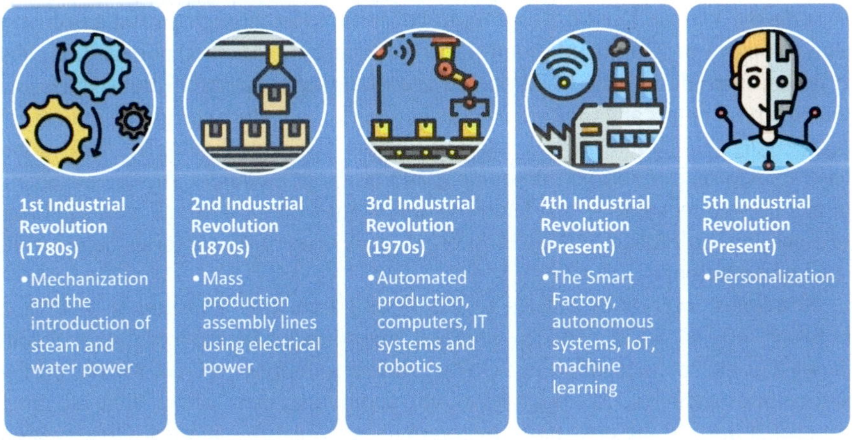

Fig. 1.6 From Industry 1.0 to Industry 5.0

- the Middle Ages: the development of craft workshops and guilds. Water mills and other mechanisms that used natural energy sources (e.g., water wheels, etc.) became more common;
- the Renaissance: engineering achievements of Leonardo da Vinci and other inventors who created drawings and models of complex mechanisms that contributed to the further development of technology.

The key features of industrial revolutions are shown in Fig. 1.6 and briefly outlined below.

The first significant shift came during the First Industrial Revolution (Industry 1.0) in the eighteenth century. Process invention was developed into automotive manufacturing instead of items produced by basic means. Starting in England in 1760, machine production reached the USA by the end of the eighteenth century. The First Industrial Revolution marked a shift from an agrarian and handicraft economy to one dominated by machinery and significantly impacted industries like mining, textiles, glass, and agriculture.

The next shift in manufacturing occurred between 1871 and 1914, known as the Second Industrial Revolution (Industry 2.0), due to extensive railroad and telegraph networks providing faster people and ideas transfer. The invention of electricity allowed factories to develop modern production lines. Thus, the first assembly line was patented in 1901 by Ransom E. Olds, producer of Oldsmobile cars. His method ensured that his company produced 20 units daily, increasing the output by 500% in one year. Consequently, Oldsmobile was creating more vehicles, resulting in a drastic price decrease at the same time. Later, Henry Ford created his system, taking Olds's method as a model. Nowadays, Ford is credited as the father of the assembly line and automotive mass manufacturing. The Second Industrial Revolution was a period of significant economic growth, with an increase in productivity, but it also caused a surge in unemployment since machines replaced many factory workers.

Industry 3.0, the Digital Revolution, began in 1969 through partial automation using memory-programmable controls and computers. The central point of this phase is the mass production and widespread use of digital logic, MOS transistors, integrated circuit chips, and their derived technologies, including computers, microprocessors, digital cellular phones, and the Internet. These technological innovations have transformed traditional production and business methods. The digital revolution converted technology that had been analog into a digital format. It is important to mention that Industry 3.0 is still present, with most of the factories being at this level of evolution.

Nowadays, everybody relates to the Fourth Industrial Revolution, known as Industry 4.0—a union between physical assets and advanced digital technologies—like the Internet of Things (IoT), artificial intelligence (AI), robots, drones, autonomous vehicles, 3D printing, cloud computing, and others, that are interconnected, having the possibility to communicate, analyze, and act. Organizations adopting Industry 4.0 are more flexible, responsive, intelligent, and, therefore, more prepared for data-driven decisions.

The term Industry 5.0 refers to people working alongside robots and smart machines. It is about robots helping humans work better and faster by leveraging advanced technologies like the Internet of Things (IoT) and big data. It adds a personal human touch to the Industry 4.0 pillars of automation and efficiency.

1.3 Future Trends

Trends in modern machine building have been widely discussed and presented worldwide. In particular, the World Manufacturing Foundation has published a report (World Manufacturing Report 2022) that describes new transformations. The following six disruptive trends for the future of manufacturing were introduced as significant opportunities to deliver solutions in excellence: cognitive manufacturing, circular manufacturing, global risks-resilient manufacturing, hyper-personalized manufacturing, rapidly responsive manufacturing, and inclusive manufacturing. These trends are likely to play out simultaneously rather than alternatively.

Further data provide some understanding of the development within manufacturing areas during the following years. Based on the (Deloitte Global 2023), the focus will be on investments in robotics, data analytics, IoT, additive manufacturing, cloud computing, and artificial intelligence (Fig. 1.7).

Robotics and automation are key components of modern manufacturing, which actively develops and introduces new technologies. The benefits of robotics and automation encompass the following: increased speed of task completion and reduced downtime; reduced errors and improved accuracy of operations; optimized resource use and reduced production costs; and rapid adaptation to changes in demand and production conditions.

Robots are ideal for automating routine and repetitive tasks such as assembly, welding, painting, etc. They are highly flexible and can quickly and efficiently

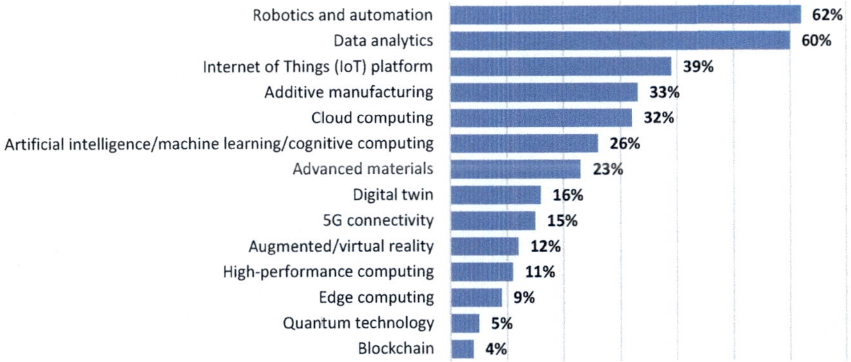

Fig. 1.7 Focus on current investments in manufacturing technologies

adapt to new production tasks when reconfiguring new production lines. Applying a modular approach to the design of flexible production lines is a relevant trend in modern production. In addition, robots are combined with traditional manufacturing processes to improve the performance of production systems and enhance product quality.

Collaborative robots (cobots) are another noteworthy feature of modern manufacturing (Pizoń et al. 2022). Cobots are easily programmable and trainable, including learning new tasks based on previous experience without manual programming and allowing the robot to adapt to new knowledge quickly. Cobots are designed to work safely with humans, with built-in sensors and security systems that allow them to work alongside employees. Modern robots are integrated with artificial intelligence (AI) for pattern recognition and data processing. This feature allows robots to recognize objects, analyze situations, and make real-time decisions.

Autonomous mobile robots powered by sensors, cameras, and AI have significant potential for autonomous navigation and obstacle avoidance in production and warehouse environments. They can quickly adapt to changes in the location and configuration of production lines or warehouses.

Using robots for automated storage and withdrawal of goods increases the speed and accuracy of order processing. This contributes to the automation of warehouse systems and the introduction of unmanned technologies in logistics and robotic warehouses.

These trends indicate a significant potential for developing robotics and automation, enabling production facilities to become more efficient, flexible, and competitive.

The digitalization of production is a leading trend in the modern industry (Adamczak et al. 2023). It encompasses the broad application of digital technologies to improve the efficiency, productivity, and flexibility of production processes. Modern digital technologies, such as the Internet of Things (IoT), artificial intelligence (AI), big data, cloud computing, and augmented reality (AR), are becoming essential tools for businesses (Haidabrus et al. 2020). They help automate production processes

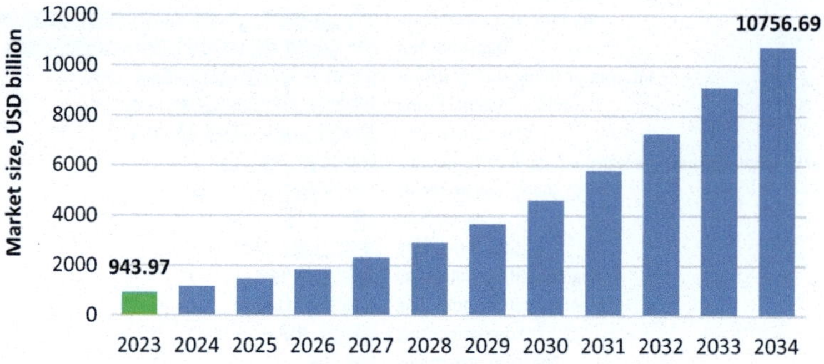

Fig. 1.8 Digital transformation market size worldwide

(Kaščak et al. 2022), reduce costs and improve efficiency (Basova et al. 2023), and improve product quality (Wieczorowski et al. 2023). Forecast data on the development of digitalization are shown in Fig. 1.8. According to forecasts, the digital transformation market will grow more than 11 times and reach USD 10,756.69 billion in 2034 (Precedence Research 2024a).

The Internet of Things (IoT) connects physical devices to the Internet to collect and exchange data, significantly improving the control and management of production processes and increasing energy efficiency (Kiyko et al. 2020). For example, temperature, humidity, and pressure sensors monitor production facilities and equipment conditions. Vibration sensors help detect potential equipment problems, such as wear or malfunctions, at early stages. Energy consumption sensors monitor the power consumption of equipment to help optimize energy costs. The information received from the sensors is transmitted to connected devices. They continuously monitor production lines and equipment and the course of production processes, analyze productivity, and identify deviations from planned indicators, ensuring the quality level (Wieczorowski and Trojanowska 2023). These inventions optimize production processes and manage equipment in real time (Haidabrus et al. 2019). IoT implementation in manufacturing significantly increases companies' competitiveness, reduces costs, and improves product quality through more accurate control and management of production processes.

The AI market is expected to continue its rapid growth and reach a value of USD 2575.16 billion by 2032 (Precedence Research 2024b). The prediction data are presented in Fig. 1.9. The growth is fueled by the increasing integration of AI into everyday technology and ongoing advances in machine learning, neural networks, and quantum computing. Breakthroughs in natural language processing, computer vision, and general AI will open new opportunities for innovation and applications.

Certain constraints or challenges, however, are also present: ethical issues (data privacy, algorithmic bias, and job displacement); regulation (the need to develop a comprehensive framework to regulate the development and use of AI for security purposes); talent shortages (demand for skilled AI professionals will exceed

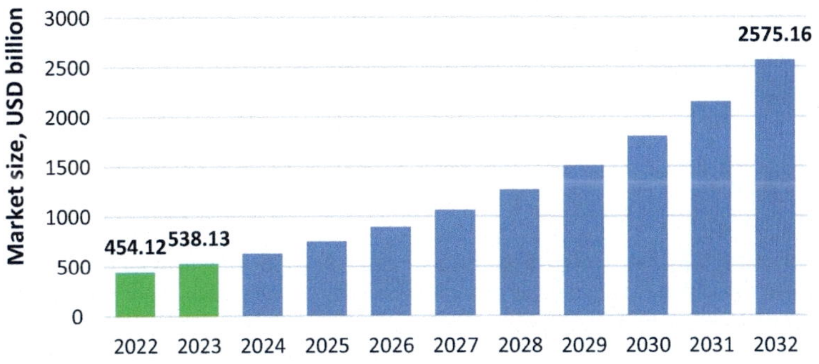

Fig. 1.9 Artificial intelligence market size worldwide

supply, requiring significant investment in education and training); and investment and funding (stimulating start-ups and research initiatives in the AI field).

Overall, the AI market is on the cusp of significant growth and transformation, impacting various industries and everyday life through continued innovation and adoption (Haidabrus 2024).

The digitalization of production has benefits that highlight its importance. Big data and analytical tools enable businesses to forecast demand, optimize supply chains, and respond quickly to market changes. Digital technologies help to quickly adapt production to changing consumer needs, introducing new products and modifications with minimal time and resources (Luscinski and Ivanov 2020). Real-time monitoring and control systems ensure high product quality by detecting defects at early stages of production and reducing waste. Digitalization also affects the labor market, creating new professions and skills requirements that lead to the reshaping of engineering education (Bilous et al. 2020; Bun et al. 2018; Ivanov et al. 2018). It requires an increase in technical literacy and knowledge of information technology (Kiyko et al. 2014). Workers must learn new skills, such as programming, automated systems management, data analytics, and agile framework (Haidabrus et al. 2023).

However, digitalization also poses specific challenges. Introducing digital technologies requires significant upfront investment, making it difficult for small and medium-sized enterprises to afford. The growing number of connected devices and systems increases the risk of cyber threats and information leaks. The transition to digital technologies may lead to job losses in traditional manufacturing sectors, requiring labor market adaptation and retraining of employees.

The digitalization of production is vital for maintaining competitiveness in the global market. It opens new opportunities for business development, efficiency improvement, and the creation of innovative products. However, to successfully implement digital transformation, businesses must consider and prepare for all potential challenges in advance.

Future trends in manufacturing are shaped by ongoing technological advancements, changing consumer demands, and a focus on sustainability (Jasiulewicz-Kaczmarek et al. 2023). Manufacturers are expected to leverage these trends to stay competitive, respond to market demands swiftly, and contribute to a more sustainable and technologically advanced industrial landscape.

References

Adamczak M, Kolinski A, Trojanowska J, Husár J (2023) Digitalization trend and its influence on the development of the operational process in production companies. Appl Sci 13:1393. https://doi.org/10.3390/app13031393

America's Investment in the Future (2000) National science foundation. USA, Arlington. https://www.nsf.gov/about/history/nsf0050/pdf/aif.pdf

Basova Y, Dobrotvorskiy S, Balog M, Iakovets A, Chelabi MA, Zinchenko A (2023) Increasing SME supply chain resilience in the face of rapidly changing demand with 3D model visualisation. Int J Mechatron Appl Mech 14:35–47. https://doi.org/10.17683/ijomam/issue14.5

Bilous O, Hovorun T, Berladir K, Dunaeva M (2021) Ensuring the Quality of Training Engineers in a Virtual Environment. In: Tonkonogyi V (eds) Advanced manufacturing processes II. InterPartner 2020. Lecture notes in mechanical engineering, Springer, Cham, pp 765–774. https://doi.org/10.1007/978-3-030-68014-5_74

Bun P, Trojanowska J, Ivanov V, Pavlenko I (2018) The use of virtual reality training application to increase the effectiveness of workshops in the field of lean manufacturing. In: 4th international conference of the virtual and augmented reality in education, VARE, pp 65–71

Cambridge University Press & Assessment (2024) Manufacturing. Retrieved from https://dictionary.cambridge.org/dictionary/english/manufacturing

CIRP Encyclopedia of Production Engineering (2014) The international academy for production engineering. In: Laperrière L, Reinhart G (eds), 1st edn. Springer Berlin, Heidelberg. https://doi.org/10.1007/978-3-642-20617-7_6561

Collins Dictionary (2024) Manufacturing. Retrieved from https://www.collinsdictionary.com/dictionary/english/manufacturing

Components Engine (2024) Manufacturing. Retrieved from https://www.componentsengine.com/site/page.wplus/Electronic-parts-catalog/English/ID_COUNT/glos_M/LN/2/wx/glos_M.html

Corporate Finance Institute (2024) Manufacturing. Retrieved from https://corporatefinanceinstitute.com/resources/valuation/manufacturing/

Degree Property & Engineering Services PVT LTD (2024) Manufacturing. Retrieved from https://www.360degreepes.com/manufacturing/

Deloitte Global (2023) Manufacturing industry outlook. Deloitte Development LLC. https://www2.deloitte.com/content/dam/Deloitte/us/Documents/energyresources/us-2023-outlook-manufacturing.pdf.

Haidabrus B (2024) Generative AI in Agile, Project, and Delivery Management. In: Ivanov V, Trojanowska J, Pavlenko I, Rauch E, Piteľ J (eds) Advances in design, simulation and manufacturing VII. DSMIE 2024. Lecture notes in mechanical engineering. Springer, Cham, pp 100–110. https://doi.org/10.1007/978-3-031-61797-3_9

Haidabrus B, Druzhinin E, Elg M, Jason M, Grabis J (2020) Programs to boost IT-readiness of the machine building enterprises. In: Ivanov V (ed) Advances in design, simulation and manufacturing II. DSMIE 2019. Lecture notes in mechanical engineering. Springer, Cham, pp 75–84. https://doi.org/10.1007/978-3-030-22365-6_8

Haidabrus B, Protsenko S, Rosenberger P, Grabis J (2020) Data analysis of readiness programs of machine-building enterprises. In: Ivanov V, Trojanowska J, Pavlenko I, Zajac J, Peraković D (eds) Advances in design, simulation and manufacturing III. DSMIE 2020. Lecture notes in mechanical engineering, Springer, Cham, pp 128–136. https://doi.org/10.1007/978-3-030-50794-7_13

Haidabrus B, Grabis J, Psarov O, Druzhinin E (2023) Agile framework as a key to information management systems delivery. In: Ivanov V, Trojanowska J, Pavlenko I, Rauch E, Piteľ J (eds) Advances in design, simulation and manufacturing VI. DSMIE 2023. Lecture notes in mechanical engineering. Springer, Cham, pp 113–120. https://doi.org/10.1007/978-3-031-32767-4_11

International Labour Organization (2022) ILO modelled estimates database, ILOSTAT SDG indicator 9.2.2—manufacturing employment as a proportion of total employment (%). Retrieved from: https://www.ilo.org/shinyapps/bulkexplorer8/?lang=en&segment=indicator&id=SDG_0922_NOC_RT_A

Ivanov V, Pavlenko I, Trojanowska J, Zuban Y, Samokhvalov D, Bun P (2018) Using the augmented reality for training engineering students. In: 4th international conference of the virtual and augmented reality in education, VARE, pp 57–64

Jasiulewicz-Kaczmarek M, Antosz K, Zhang C, Ivanov V (2023) Industry 4.0 technologies for sustainable asset life cycle management. Sustainability 15:5833. https://doi.org/10.3390/su15075833

Kaščak J, Husár J, Knapčíková L, Trojanowska J, Ivanov V (2022) Conceptual use of augmented reality in the maintenance of manufacturing facilities. In: Trojanowska J, Kujawińska A, Machado J, Pavlenko I (eds) Advances in manufacturing III. Manufacturing 2022. Lecture notes in mechanical engineering, Springer, Cham, pp 241–252. https://doi.org/10.1007/978-3-030-99310-8_19

Katina PF, Cash CT, Caldwell LR, Beck CM, Katina JJ (2023) Advanced manufacturing management: a systematic literature review. Sustainability 15(6):4702. https://doi.org/10.3390/su15064702

Kenton W (2024) Manufacturing: definition, types, examples, and use as indicator. Investopedia. Retrieved from https://www.investopedia.com/terms/m/manufacturing.asp

Kiyko SG, Druzhinin EA, Koba SA, Haidabrus BV (2014) A mathematical background for information technology of project's processes integration taking into account risk factors. Metallurgic Minin Industr 6(3):66–70

Kiyko S, Druzhinin E, Prokhorov O, Haidabrus B (2020) Multi-agent model of energy consumption at the metallurgical enterprise. In: Ivanov V, Trojanowska J, Pavlenko I, Zajac J, Peraković D (eds) Advances in design, simulation and manufacturing III. DSMIE 2020. Lecture notes in mechanical engineering, Springer, Cham, pp 156–165. https://doi.org/10.1007/978-3-030-50794-7_16

Levinson M (2017) What is manufacturing? why does the definition matter? Congress Res Serv. https://www.nist.gov/system/files/documents/2017/02/08/r44755.pdf#:~:text=URL%3A%20https%3A%2F%2Fwww.nist.gov%2Fdocument%2Fr44755pdf%0AVisible%3A%200%25%20

Luscinski S, Ivanov V (2020) A Simulation study of industry 4.0 factories based on the ontology on flexibility with using FlexSim® software. Manag Product Eng Rev 11(3):74–83. https://doi.org/10.24425/mper.2020.134934

Merriam-Webster's Collegiate Dictionary (2003) 11th edn. by Frederick C. Mish. Merriam-Webster. 978–0–87779–471–4

Mishra R, Pundir AK, Ganapathy L (2014) Manufacturing flexibility research: a review of literature and agenda for future research. Glob J Flex Syst Manag 15:101–112. https://doi.org/10.1007/s40171-013-0057-2

Muzylyov D, Shramenko N (2020) Blockchain technology in transportation as a part of the efficiency in industry 4.0 strategy. In: Tonkonogyi V (eds) Advanced Manufacturing Processes. InterPartner 2019. Lecture notes in mechanical engineering, Springer, Cham, pp 216–225. https://doi.org/10.1007/978-3-030-40724-7_22

Novinkina J, Davydovitch A, Vasiljeva T, Haidabrus B (2021) Industries pioneering blockchain technology for electronic data interchange. Acta Logistica 8(4):319–327. https://doi.org/10.22306/al.v8i4.230

Oxford English Dictionary (2010) 3rd edn. Oxford University Press, New York. 978–0199571123

Oxford Learner's Dictionary (2024) Manufacturing. Retrieved from https://www.oxfordlearnersdictionaries.com/definition/english/manufacturing

Pizoń J, Gola A, Świć A (2022) The role and meaning of the digital twin technology in the process of implementing intelligent collaborative robots. In: Gapiński B, Ciszak O, Ivanov V (eds) Advances in manufacturing III. MANUFACTURING. Lecture notes in mechanical engineering, Springer, Cham, pp 39–49. https://doi.org/10.1007/978-3-031-00805-4_4

Precedence Research (2024a) Digital Transformation market size and growth 2024 to 2033. Retrieved from https://www.precedenceresearch.com/digital-transformation-market

Precedence Research (2024b) Artificial intelligence (AI) market size, share, and trends 2024 to 2034. Retrieved from https://www.precedenceresearch.com/artificial-intelligence-market

Relich M, Nielsen I, Gola A (2022) Reducing the total product cost at the product design stage. Appl Sci 12:1921. https://doi.org/10.3390/app12041921

Srivatsan TS, Manigandan K, Sudarshanet TS (2018) Manufacturing techniques for materials, 1st edn. CRC Press, Boca Raton. https://doi.org/10.1201/b22313

United Nations Economic Commission for Europe (2024) INDICATOR 9.2.2. Manufacturing employment as a proportion of total employment. Retrieved from https://w3.unece.org/SDG/en/Indicator?id=166

Wieczorowski M, Trojanowska J (2023) Towards metrology 4.0 in dimensional measurements. J Mach Eng 23(1):100–113. https://doi.org/10.36897/jme/161717

Wieczorowski M, Trojanowska J, Sokolov O (2023) Digital and information technologies in metrology 4.0. In: Ivanov V, Trojanowska J, Pavlenko I, Rauch E, Piteľ J (eds) Advances in design, simulation and manufacturing VI. DSMIE 2023. Lecture notes in mechanical engineering, Springer, Cham, pp 81–90. https://doi.org/10.1007/978-3-031-32767-4_8

World Bank (2022) World Development Report: finance for an equitable recovery. World Bank Group, Washington. https://doi.org/10.1596/978-1-4648-1730-4

World Manufacturing Report (2022) Redesigning supply chains in the new era of manufacturing. World Manufacturing Foundation, Italy, Milano. 978–88–943861–9–6

WTO: World Trade Report (2021) Economic resilience and trade. World Trade Organization, Geneva. 978–92–870–5140–0

Chapter 2
Overview of Manufacturing

2.1 Types of Products

Engineering output usually concerns machines. A machine is a mechanism or a combination of mechanisms that perform work or convert energy (Zakharkin 2004). There are enterprises that manufacture a large engineering output (e.g., pistons, ball, and roller bearings). Such an output is called a product. Depending on their use purpose, products can be obtained from main and auxiliary manufacture. The former is realized as goods. The latter is consumed within the enterprise itself (products of tool workshops, dedicated technological equipment, etc.). According to the valid standards, there are such types of products: part, assembly unit, complex, and kit.

Part is a homogeneous product manufactured without assembly processes. It can be covered with electrodeposit or paint-and-lacquer coatings.

An assembly unit is a product part made separately and used for further manufacturing processes.

Complex is two or more specific non-assembly products applied for interdependent operation (CNC machine, robotic unit, etc.).

Kit is two or more non-assembly products manufactured jointly for auxiliary use (kit of wrenches, machine spare parts, etc.).

As an end production item, each machine may be subdivided into smaller products—assembly units and parts. Its structure is shown in Fig. 2.1. Parts may belong to products on any subdivision level.

A correct machine subdivision into units of different levels leads to the proper assembly process and determines the sequence of part manufacturing. In other words, machines are produced reasonably. Moreover, units may be manufactured simultaneously, which raises manpower productivity.

Let us observe a double-stage reducer as a product (Fig. 2.2) and analyze its specification (Table 2.1). Reducers are applied to convert torques of electric motors or other power machines. They are used to decrease the motor rotational rate and increase output torques (Krol et al. 2023). The cylindrical reducer is most widely

V. Ivanov et al., *Fundamentals of Manufacturing Engineering Using Digital Visualization*, Springer Tracts in Mechanical Engineering,
https://doi.org/10.1007/978-3-031-74360-3_2

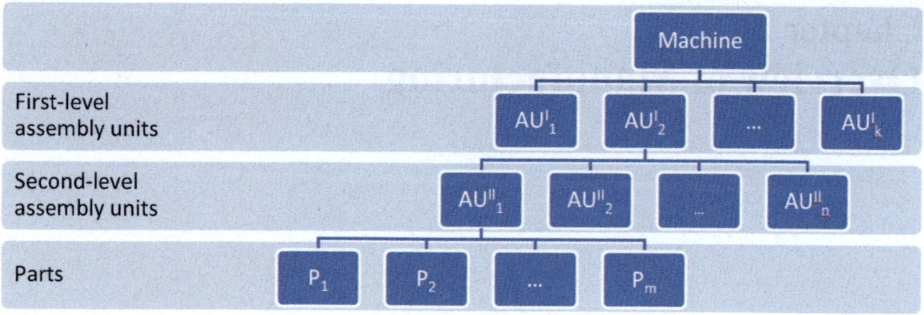

Fig. 2.1 Hierarchical product structure (a general model)

applied to convert torques. Its working principle is based on spur-gear drives. In contrast to worm reducers, such a mechanism applies only spur gears (Sokolov et al. 2020; Shevchenko et al. 2020). Traditionally, design process is realized using CAD software (Tigariev et al. 2020).

The cylindrical reducer principle: the torque within the motor and the input shaft is transferred and converted (raised) via gear drives with different teeth on the way to the output shaft. Reducer specification differs by gearing ratios. The latter depends on shaft-gear teeth. Besides, the gearing ratio depends on a number of its stages. The more stages we have, the higher the gearing ratio is with the product assembly and price.

The hierarchical structure of the reducer is shown in Fig. 2.3. The assembly process of ball bearings is not described in this case study. They are considered as finished products. The quantity of parts used in the assembly unit is mentioned in the brackets, e.g., "Reinforced cuff (2)", "Pin (2)", etc. This means that one left cover, two reinforced cuffs, and two pins will be used to assemble the second-level assembly unit ("Left cover (2nd lvl. assy. unit)").

SCAN QR
ManuVis
Manufacturing Visualization

To access the digital content in this subchapter, please follow these steps:

(1) open the ManuVis application on your device;

(2) scan the QR code

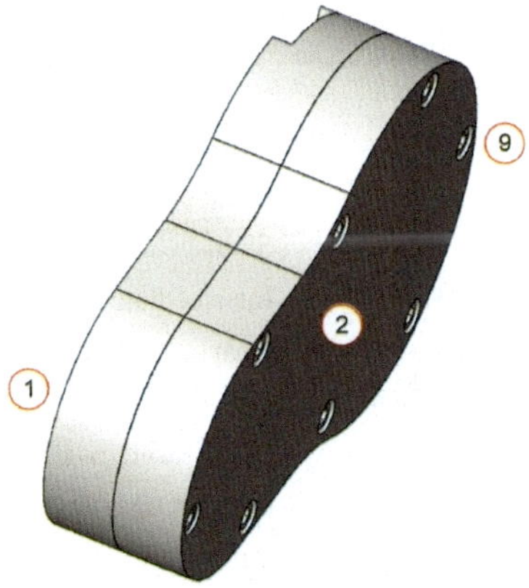

a)

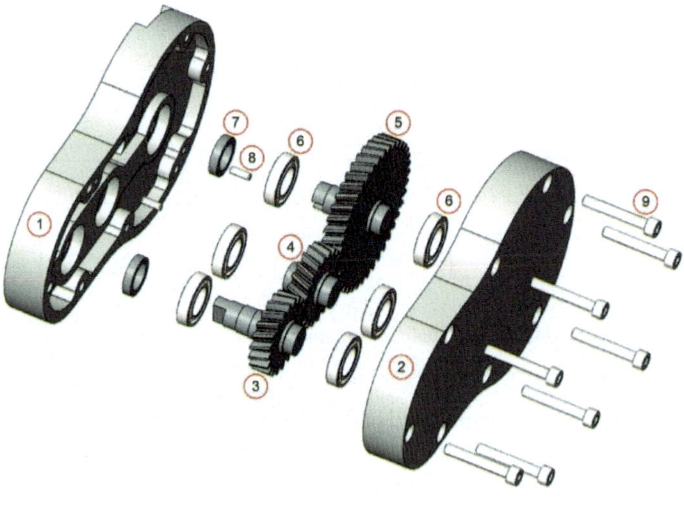

b)

1 – left cover; 2 – right cover; 3 – shaft-gear; 4 – intermediate shaft; 5 – gear; 6 – ball bearing; 7 – reinforced cuff; 8 – pin; 9 – screw

Fig. 2.2 Product "double-stage reducer": **a**—assembled view; **b**—exploded view

2.2 Technological Assembly Scheme

The product assembly sequence is conditioned by its design features and accuracy methods (Kusyi and Stupnytskyy 2020; Borucka et al. 2023). The product assembly sequence requires creating a technological assembly scheme. It simplifies assembly

Table 2.1 Part list

Item	Part name	Quantity
Parts		
1	Left cover	1
2	Right cover	1
3	Shaft-gear	1
4	Intermediate shaft	1
5	Gear	1
Standard parts		
6	Ball bearing 6–1000903	6
7	Reinforced cuff 23–16-6	2
8	Pin 7 × 30	2
9	Screw M6 × 40	8

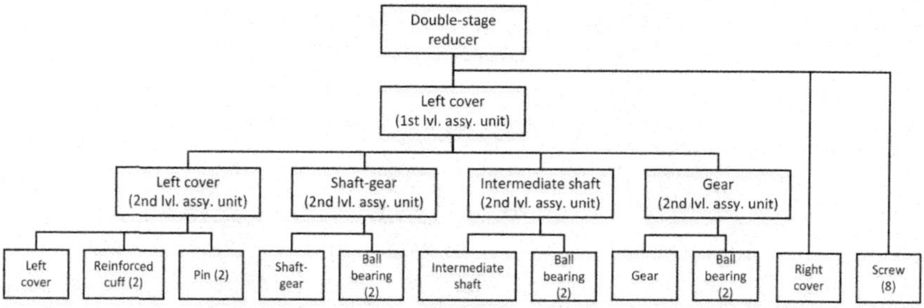

Fig. 2.3 Hierarchical double-stage reducer structure

organizing, product kitting, unit and part delivery to assembly places, and production planning.

The initial stage of developing the product assembly sequence is a study of product design, operating principles, and technical conditions of its receiving and testing. Here, we should define the aim of service product use, its general tasks, and all auxiliary quantitative requirements.

Based on the assembly design, the product assembly diagram is generated. Before drawing this diagram, the product should be disintegrated into indivisible parts for a proper assembly sequence.

It is necessary to define as many assembly units as possible. By assembling them separately and simultaneously, we can reduce the time for the assembly cycle itself. A reasonable assembly diagram is created according to Fig. 2.4. Parts and units are arranged in an appropriate sequence.

The scheme represents parts and assembly units as rectangles with three sections (Fig. 2.5). The upper section is the part, assembly unit, or product name. The lower

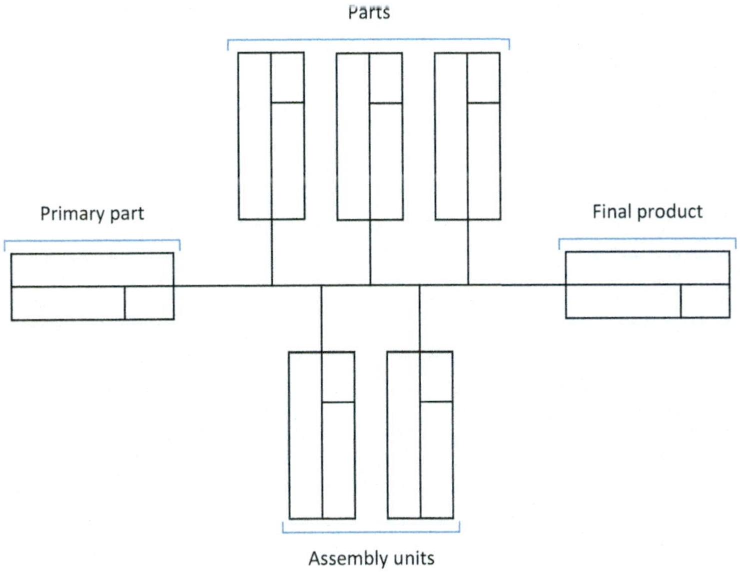

Fig. 2.4 Technological assembly scheme (general model)

left section contains the item code according to the specification. The lower right section is the quantity of parts assembled simultaneously.

Each assembly unit is assigned a number of the part of the unit to which it belongs. The initial part to start the node or product assembly is called primary. The assembly unit in the product design is the main first-level assembly unit. The assembly unit within the first-level one is the second-level assembly unit, etc.

Fig. 2.5 Example of filling in sections of the technological assembly scheme

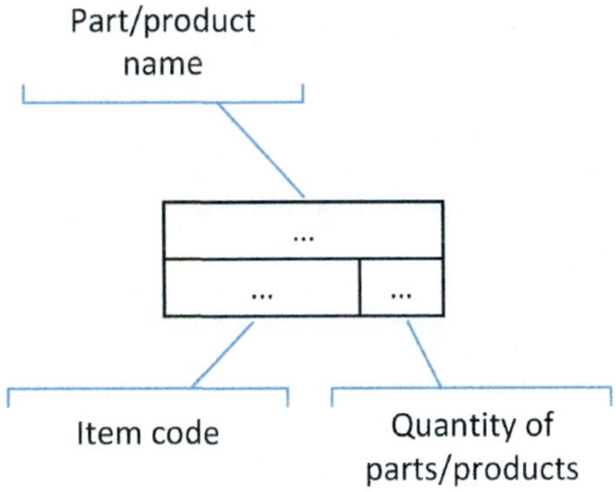

To access the digital content in this subchapter, please follow these steps:
(1) open the ManuVis application on your device;
(2) scan the QR code

SCAN QR ManuVis Manufacturing Visualization

The technological assembly scheme is built in the following way. A rectangle as a basic part is drawn with a horizontal line from it (assembly flow) to another rectangle as an end product. Above the horizontal line, arranged parts within the product assembly are shown. Under the horizontal line, the arranged assembly units are depicted. The technological assembly scheme may be enlarged and detailed depending on the product's complexity. The former is reasonable for difficult assembly cases. The latter is optimal for subunit assembly within the unit assembly itself. Also, we should note additional marks on the scheme to explain operations during the assembly process (pressing, soldering, scraping, etc.). The marks are shown near corresponding elements.

The generated technological assembly scheme is a basis for projecting the assembly process flow or cycle schemes.

For example, Fig. 2.6 represents the assembly process flow for a double-stage reducer. The primary part is the left cover (Position 1). The assembly starts with installing two reinforced cuffs (Position 7). Consequently, the assembled left cover with the cuffs is obtained. Then, two pins (7×30) are inserted into the cover. Finally, the assembled second-level cover (Fig. 2.6) is obtained.

Later, ball bearings 6–1000903 (Position 6) are pressed onto the shaft-gear (Position 3) from both sides. Therefore, we obtain the second-level assembly of the shaft-gear (Fig. 2.7).

Similarly, ball bearings 6–1000903 (Position 6) are pressed onto the intermediate shaft (Position 4) from both sides. It results in the second-level assembly of the intermediate shaft (Fig. 2.8).

Likewise, in the previous cases, ball bearing 6–1000903 (Position 6) are pressed onto the gear (Position 5) from both sides. It gives the second-level assembly of the gear (Fig. 2.9).

The assembled shaft-gear, intermediate shaft, and gear are installed into the second-level assembled left cover. It produces the first-level assembled left cover (Fig. 2.10).

Further, the right cover is installed onto the first-level assembled left cover. They are adjusted via pins and screwed by eight screws M6 $\times$ 40 (Position 9) with a torque of 30 $\mathrm{H} \cdot \mathrm{m}$ (Fig. 2.11). Preliminarily, screws are oiled with a locking varnish.

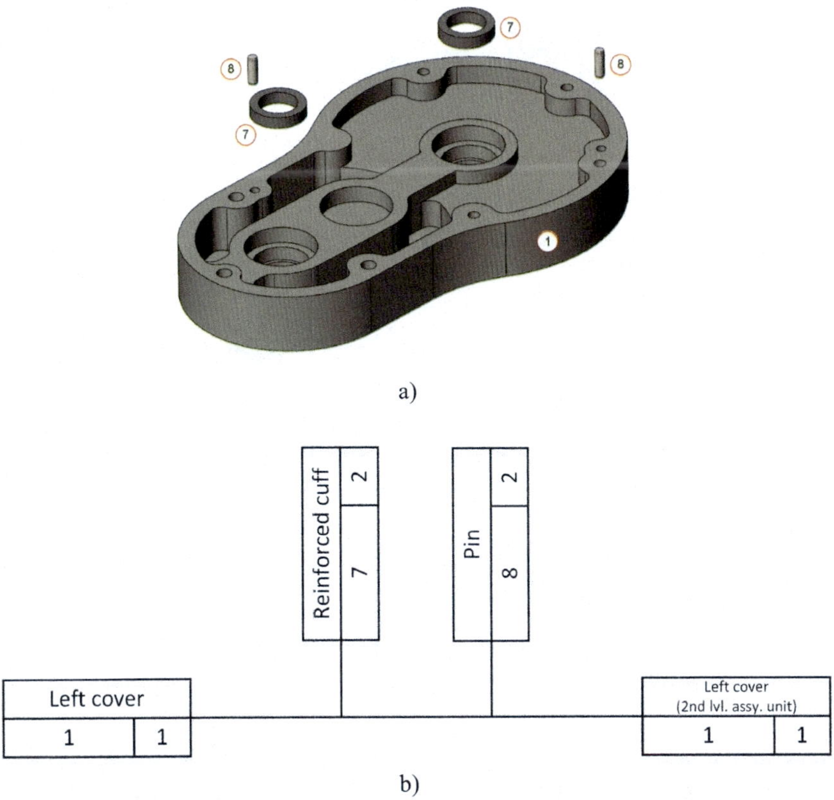

Fig. 2.6 Second-level assembly of the left cover: **a** exploded view of the assembly unit; **b** technological assembly scheme

Disassembly of the double-stage reducer is performed in the following way (Fig. 2.2). Eight screws M6 × 40 (Position 9) are removed. The right cover is removed to disengage the reducer halves. From the first-level assembled left cover, the assembled gear, intermediate shaft, and shaft-gear are taken. It leaves separately the second-level assembled left cover. Ball bearings (Position 6) are taken from the shaft-gear (Position 3), intermediate shaft (Position 4), and gear (Position 5). Two pins 7 × 30 are removed from the second-level assembled left cover. It gives separately the pins (Position 8) and the assembled left cover with the reinforced cuffs. Finally, two reinforced cuffs (Position 7) are removed from the assembled left cover. That leaves the left cover separately (Position 1).

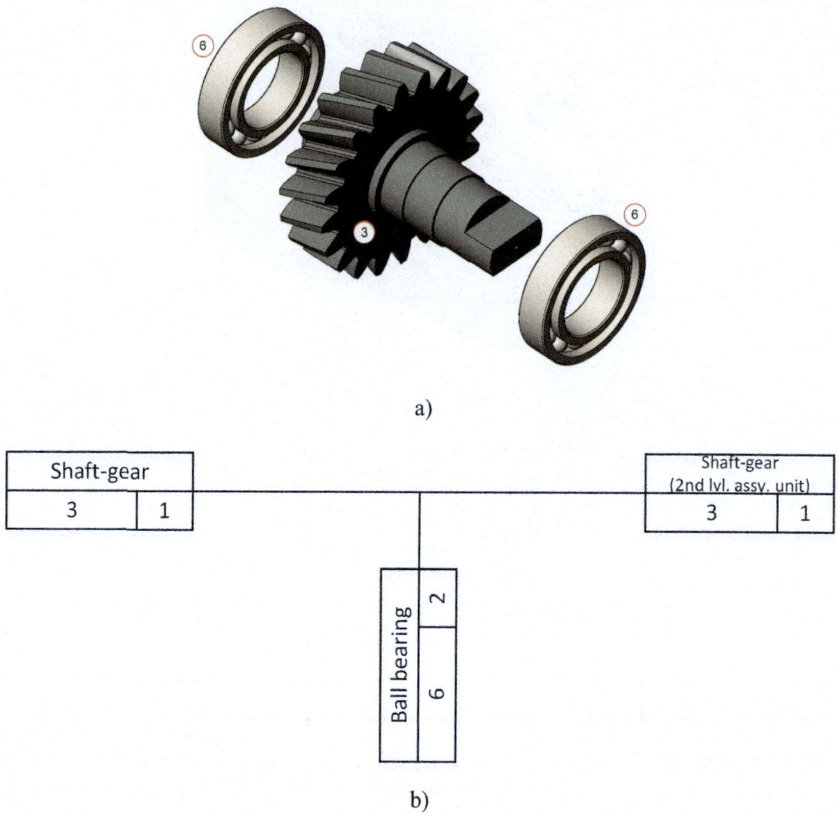

Fig. 2.7 Second-level assembly of the shaft-gear: **a** exploded view of the assembly unit; **b** technological assembly scheme

To access the digital content in this subchapter, please follow these steps:
(1) open the ManuVis application on your device;
(2) scan the QR code

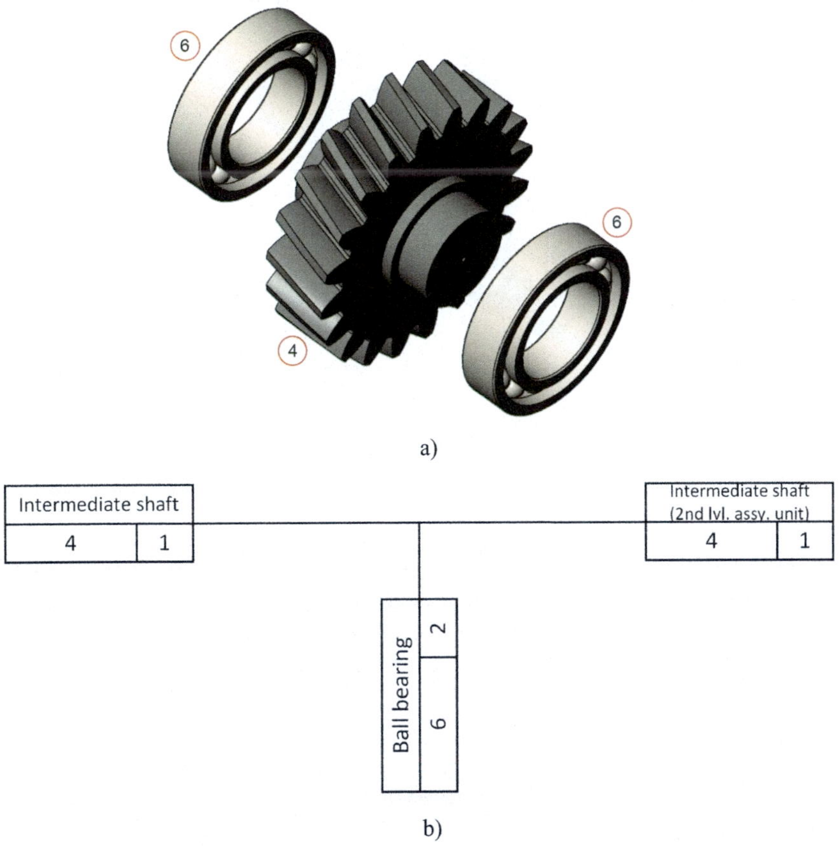

Fig. 2.8 Second-level assembly of the intermediate shaft: **a** exploded view of the assembly unit; **b** technological assembly scheme

2.3 Types of Manufacturing

All engineering enterprises are characterized by manner and type of manufacture (Szczepaniak and Trojanowska 2019). Manner of manufacture is a classifying feature to distinguish methods of producing something. For example, there can be foundry, forging, welding, machine assembling, thermal manufacture, etc.

Type of production is a classifying feature to distinguish the range, regularity, stability, and volume of made products. For example, there can be single-part, batch, and mass manufacturing. Let us analyze their main attributes.

The single-part manufacturing is characterized by a wide range of makable and repairable products. Their production volume is small.

Production volume is the quantity of products with a certain name, form, and size made or repaired by an enterprise within a time interval.

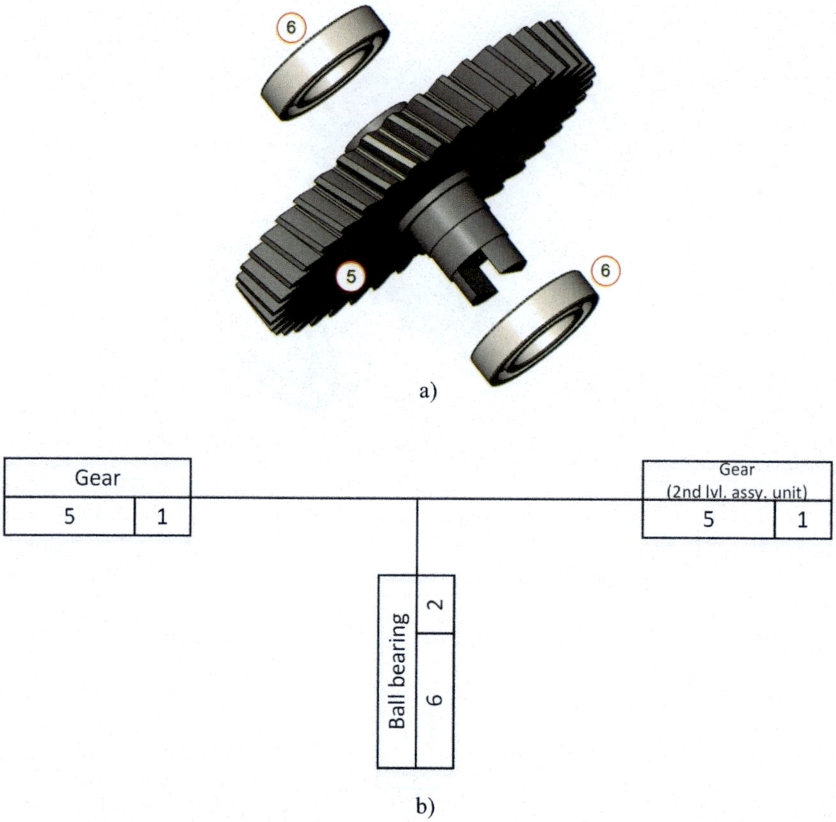

Fig. 2.9 Second-level assembly of the gear: **a** exploded view of the assembly unit; **b** technological assembly scheme

During single-part manufacturing, technological operations are performed in working areas without iterations. Universal equipment and fixtures are used (three-jaw chucks, vises, etc.) (Karpus' and Ivanov 2008). Special machines are used only in exclusive cases. Measuring and cutting tools are universal as well. Within a workshop, equipment is arranged by type (areas of lathes, milling machines, etc.). Workers are high-skilled specialists.

The batch manufacturing comprises a stabler range of products. Small-batch, medium-batch, and large-batch manufacturing can be used depending on the volume of iterating batches. The small-batch and medium-batch manufacturing may include CNC machines. The large-batch industries apply specialized (Krol et al. 2024; Kushnirov et al. 2023) and modular machines (Yakovenko et al. 2023, 2021; Zaleta et al. 2023). Fixtures are universal, modular, or adjustable (Karpus and Ivanov 2012). It is efficient to use group machining to organize object-closed areas to produce specific parts. Equipment in such areas is arranged according to the order of technological

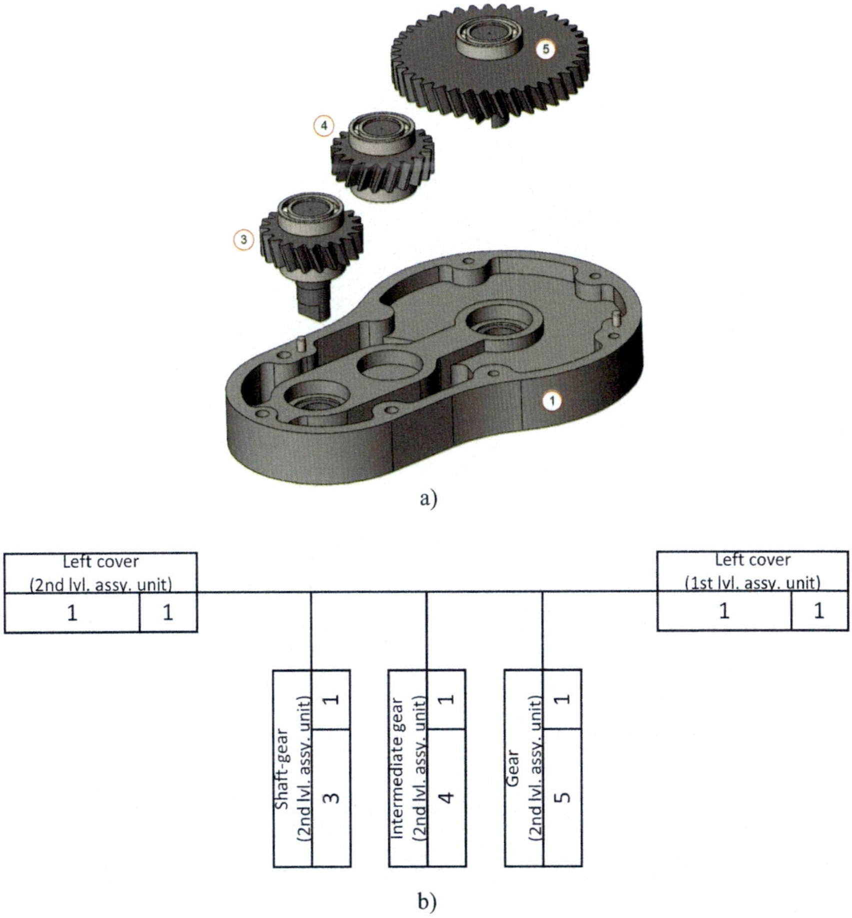

Fig. 2.10 First-level assembly of the left cover: **a** exploded view of the assembly unit; **b** technological assembly scheme

operations. Cutting tools are universal (e.g., drills, cutter bits, etc.) or specialized (core drills) (Gasanov et al. 2019). Measuring tools are universal (scaled). There can be gages, including go-no-go ones. Workers are semi-skilled specialists. Sometimes, object-closed cells are organized as the variable process flow manufacturing lines. Here, it is reasonable to determine the necessary order of machining similar parts.

The mass manufacturing permanently makes a vast number of products for a long time. To perform the same technological operation, it is necessary to conduct the work area specialization. Equipment in such areas is arranged according to the order of technological operations. There may be specialized machines, automated lines, etc. Fixtures are usually dedicated (for a particular operation). Cutting tools are specialized and combined, sometimes universal. Multiple adjustability is present

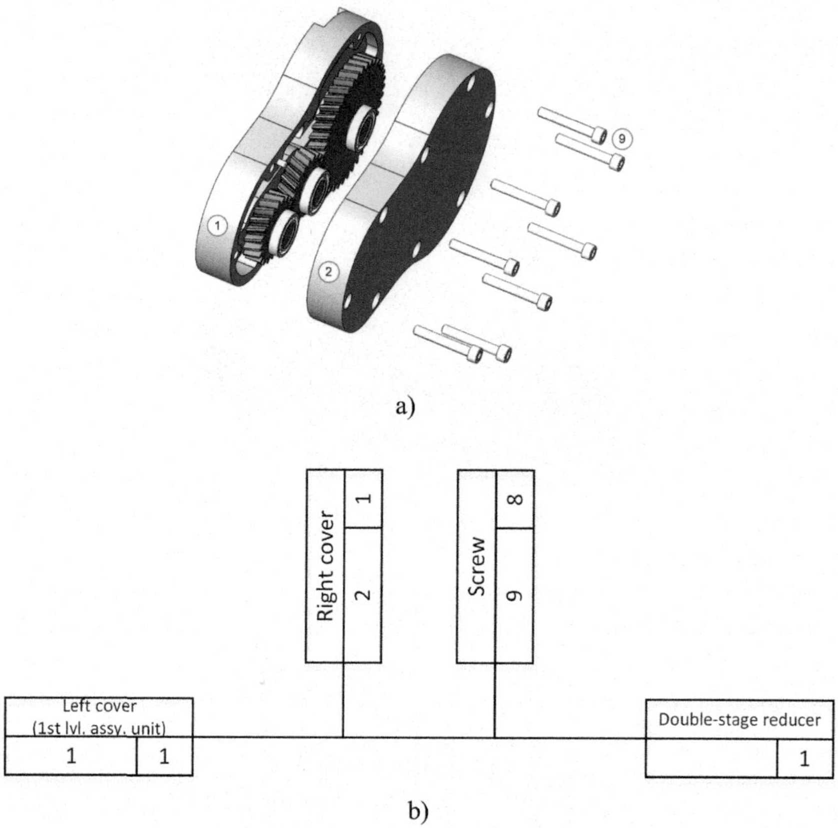

Fig. 2.11 Assembly of double-stage reducers: **a** 3D model of the product; **b** technological assembly diagram

as well. Measuring tools are gages, including go-no-go ones. Mechanized and automated transportation of workpieces (e.g., conveyor belts, gravity slides, slopes, etc.) is widely applied. Requirements for machine adjusters are high and for machine operators are low. One of the mass manufacturing attributes is manufacturing tact. It is a time interval between the production start and end of two successive products.

2.4 Technical Manufacture Preparation

Technical manufacture preparation comprises activities aimed at developing and preparing a new product. It generally includes scientific, design, technological, and organizational manufacture preparation (Varela et al. 2021).

The scientific manufacture preparation conducts research to identify possibilities of using the latest achievements of fundamental sciences (e.g., physics, chemistry, etc.) in a new product. It also promotes new operation principles of various machines. The design manufacture preparation develops drawings of assembly units and parts, various instructions for producing, operating, and repairing machines, etc. The technological manufacture preparation includes measures to prepare the industry for manufacturing or repairing new products. The main functions are testing products for manufacturability, designing technological processes, developing and producing equipment, and organizing all technological preparations (Denysenko et al. 2020). The organizational manufacture preparation concerns work on production planning, provision of component parts to make a product, etc.

2.5 Manufacturing and Technological Processes

To produce a machine that meets its intended purpose and technical requirements, the enterprise must sequentially transform workpieces into finished parts, assembly units, and machines (Kusyi et al. 2022). They are finally tested and shipped to consumers (Shramenko and Muzylyov 2019; Muzylyov et al. 2021). The order of these actions is called the manufacturing process.

The manufacturing process deals with all actions of people and tools required to make or repair products. The manufacturing process includes various technological processes: procurement (e.g., forging, casting, etc.), machining, and other operations. Therefore, the manufacturing process is a broader concept than the technological process.

The technological process is a completed stage of the manufacturing process (Kusyi et al. 2024). It comprises purposeful actions to change or determine the labor object state. In general, the technological process consists of such elements: operations, setups, positions, steps, passes, and techniques (Fig. 2.12).

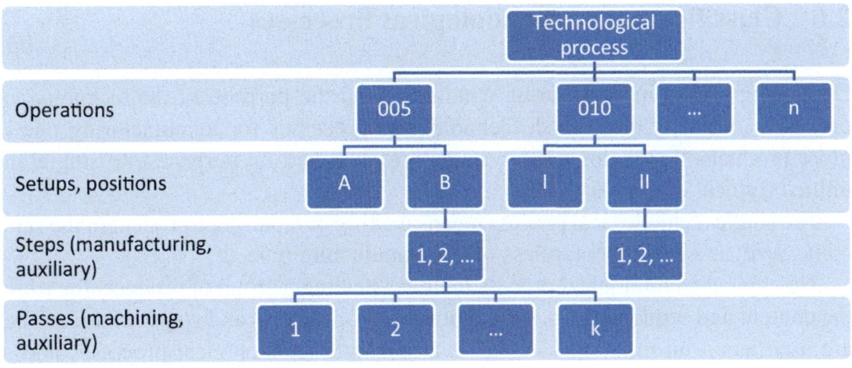

Fig. 2.12 Technological process structure

Operation is a completed stage of the technological process performed within one working area. Operations are numbered one in five (005, 010, …).

Setup is a stage of technological operation carried out during the permanent fixing of a workpiece or assembly unit. In drawings, they are denoted by capital letters (A, B, …).

Position is a location occupied by a permanently fixed workpiece or unit during assembly together with a fixture relative to a tool or fixed part of the equipment when the operation stage is performed. Positions are indicated by Roman numerals (I, II, …).

When multi-spindle equipment is used according to the two-index scheme (turning the table to the alternate rather than adjacent position: I to III, etc.), the A setup for even positions and the B setup for odd positions can be implemented. The fixtures for locating and clamping workpieces on the A and B setups can differ.

The manufacturing step is a completed stage of a technological operation characterized by using the same equipment at constant technological modes and unchanged fixturing.

The auxiliary step is a completed stage of a technological operation consisting of human (or equipment) actions that are not accompanied by changes in the properties of labor items but are necessary to perform a manufacturing step.

The machining pass is a completed stage of a manufacturing step that comprises a single movement of a tool to a workpiece, accompanied by changes in shape, size, surface quality, and properties of the workpiece.

The auxiliary pass is a completed stage of a manufacturing step that concerns a single movement of a tool to a workpiece, which is not accompanied by changes in shape, size, surface quality, and properties of the workpiece but is necessary for preparing a machining pass.

A technique is a completed set of human actions used to conduct a step or stage for one purpose. It deals with taking a workpiece, setting it up in a fixture, turning on a machine, etc.).

2.6 Classification of Technological Processes

Depending on the manufacturing conditions and the purpose of the technological processes, we may distinguish technological processes for manufacturing one or more products. Technological processes are classified by purpose into single and unified (typical or group).

The single technological process means making or repairing a product of the same name, size, and design, regardless of the manufacture type.

The typical technological process includes making a group of products for which the content and sequence of most technological operations and steps coincide. They are used as an information base to develop single technological processes and as standards for typical technological processes.

The typification of technological processes is based on part classification based on the commonality of configuration and similarity of technological processes. For example, the following classes of parts are distinguished: shafts, axles, bushings, disks, plates, frames, etc. Typification is reasonable for generalizing the existing advanced technological processes and sharing the experience of introducing new equipment and tools. This idea has been implemented at many enterprises.

The group technological process makes a group of products with different designs but common technological features. The group technology develops typification. It promotes manufacturing or assembly, dramatically reducing the time required to change equipment. Group technology is also based on classifying products and assembling groups. However, the structural similarity of products is a secondary attribute. With group technology, the technological process is designed for a complex part, either the most complex part of the group that actually exists or artificially made as a part containing all surfaces of the individual group parts.

The technological process developed for a complex part is usually redundant in the case of specific parts. It may contain technological operations and steps to machine absent surfaces. Via the group technological process, the single technological processes are developed by excluding unnecessary operations and steps and specifying equipment. This principle is a basis for one of CAPP's directions (designing single technological processes based on a unified one).

As to the level of science and technology achievements, technological processes can be divided into working and promising ones.

The working technological process is performed according to working documentation that reflects the capabilities of a particular manufacturing company.

The promising technological process corresponds to technical solutions that the enterprise should fully or partially implement (e.g., new machines, processing methods, equipment, etc.).

The temporary technological process is used at the enterprise for a limited period because of equipment repair or accidents.

The complex technological process includes technological operations and moving, controlling, cleaning workpieces, etc.

All of the above-mentioned technological processes can be developed with various technical solutions. Depending on this, technological processes are recorded on various forms of technological documentation.

2.7 Production Lines for Machining and Assembly

The forms and principles of machining and assembly organization significantly depend on the manufacture type (Jasiulewicz-Kaczmarek et al. 2023; Grześkowiak and Trojanowska 2020). For example, single-part and small-batch manufacturing conditions require the arrangement of equipment in workshops by type of work (e.g., areas of lathes, milling machines, etc.).

The medium-batch industry applies group machining with the organization of object-closed areas to produce parts (structurally and technologically similar). Equipment in such areas is arranged by order of group technological operations. Such an area is constantly adjusted to produce any part of the group. The quantity of parts in a group can reach dozens or even hundreds of items. It can dramatically reduce equipment changeover time and use more efficient machines in medium-batch manufacturing.

Usually, process flow lines are organized in manufacturing. There can be:

- the continuous process flow lines: operations are simultaneous, equal, or multiple of tact (workpieces are transported and treated by and between machines constantly);
- the direct process flow lines: operations are non-simultaneous and completed in technological sequence (workpieces may not be treated between machines).

Both process flow lines can be single-nomenclature and multi-nomenclature (with the manufacture of several products in a certain sequence).

The choice of the form of technological process organization depends on the manufacture type, assembly complexity, product design, and size (Jasiulewicz-Kaczmarek et al. 2022). At the first classification level, the following three organizational assembly forms can be distinguished (Fig. 2.13).

The non-flow assembly can be as follows (Fig. 2.14). The non-flow stationary assembly is characterized by the product remaining at the same workplace throughout the assembly process, with the basic part installed on the stand or floor. All parts, assembly units, and components are delivered to this workplace.

This type can be performed without dismemberment when the entire assembly is done by one worker or team from start to finish (namely, there is a general assembly). The foreman assigns individual assembly steps to workers if a team carries out the assembly. Such an approach is used in single-part and small-batch manufacturing to assemble a small product by one worker. Besides, it is applicable to assemble medium and large products by a team in single-part manufacturing.

The non-flow stationary assembly with dismemberment implies the presence of a subassembly (in addition to the general assembly). In this case, units of the first,

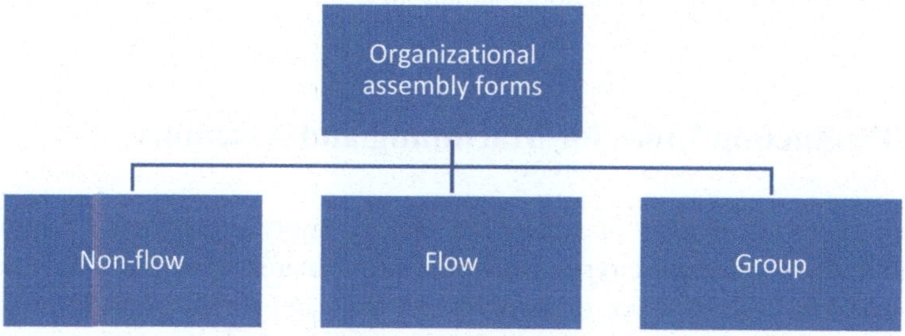

Fig. 2.13 Classification of the organizational assembly forms

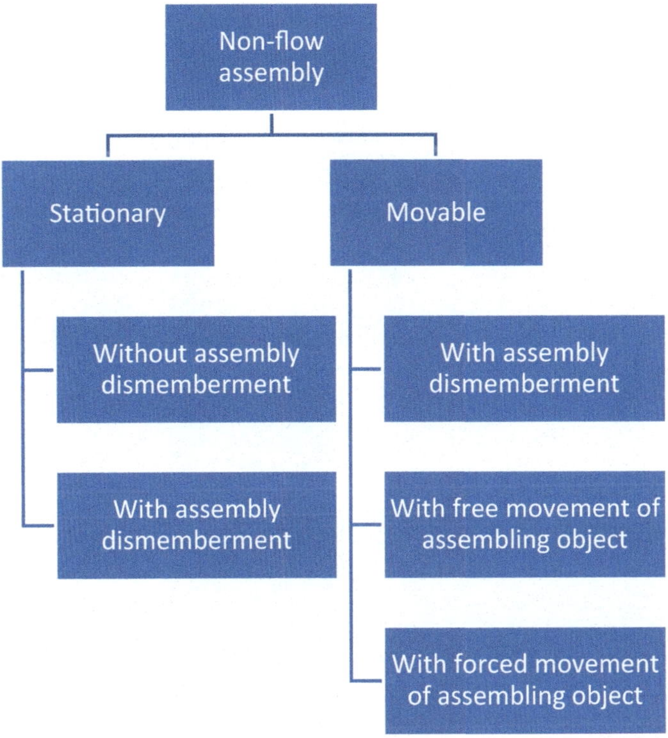

Fig. 2.14 Types of non-flow assembly

second, and other orders are assembled simultaneously by different workers or entire teams. This type is appropriate for assembling medium and large products in small-batch manufacturing.

The non-flow movable assembly is characterized by the successive movement of the product being assembled within several work areas. The technological process is divided into operations performed by one worker or a small team.

As the operation is being conducted, the product can move to the next work area manually or using transport equipment.

The forced movement assembly means that the product moves via a conveyor or a trolley inside a closed transport system with a drive chain. The non-flow movable assembly is reasonable in medium-batch manufacturing.

The flow assembly always involves work dismemberment. It can be implemented in the following variants (Fig. 2.15).

The stationary flow assembly is used for large and bulky products, such as heavy machines, marine diesel engines, and airplanes. All products are assembled in one place. Workers or teams simultaneously move from one assembly object to the next in time periods equal to a tact. Workers or a team at each of the assembly

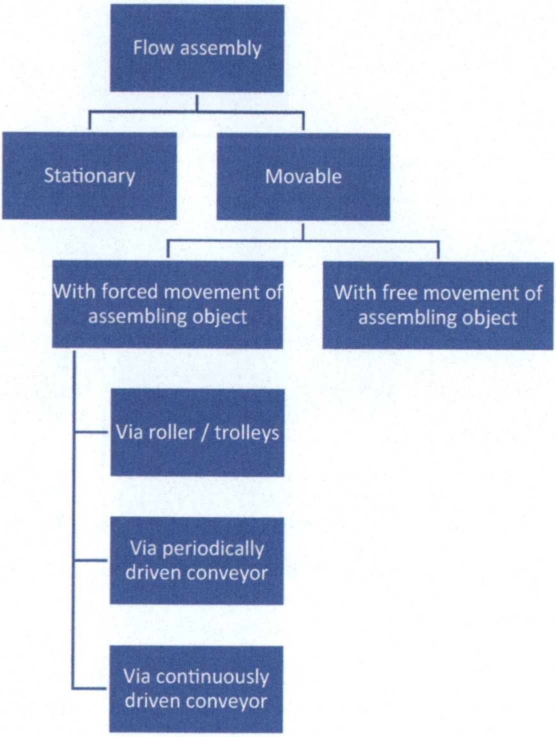

Fig. 2.15 Types of flow assembly

objects performs the same operation. The scope of application is the medium-batch manufacturing of large products.

The movable flow assembly is reasonable in the large-batch and mass manufacturing. As a rule, the duration of assembly operations is equal to or a multiple of a tact. Several assemblers work simultaneously on longer operations.

The flow assembly can be organized with free or forced movement of the object. In the first case, the worker transfers the assembling product to the next operation after completing his workload. In the second case, the moment of product movement is determined by a signal (light or sound), after which the forced periodic movement of the conveyor begins. Sometimes, the conveyor moves continuously when the operation is being conducted.

The inter-operational movement of the assembling product is carried out:

- manually or via trolleys, inclined trays, or rollers;
- via periodically driven conveyor; during movement of which assembly operations are performed;
- via continuously driven conveyor at a speed which ensures that assembly operations are possible (0.25–3.5 m/min).

The above-mentioned organizational assembly forms are summarized in Table 2.2.

Table 2.2 Organizational assembly forms

Features defining the organizational assembly form		Type of manufacturing														
		Single part			Small batch			Medium batch			Large batch			Mass		
Size of products	Small	+			+			+			+			+		
	Medium		+			+			+			+			+	
	Large			+			+			+			+			+
Flow assembly											+	+	+	+	+	+
Non-flow assembly		+	+	+	+	+	+	+	+	+						
Stationary assembly		+	+	+	+	+	+					+	+		+	+
Movable assembly								+	+	+	+	+	+	+	+	
Without assembly dismemberment		+	+	+	+											
With assembly dismemberment			+	+		+	+	+	+	+	+	+	+	+	+	+
With free movement of assembling object								+	+							
With a forced movement of assembling object											+	+		+	+	
Transfer of assembling teams										+			+			+

References

Borucka A, Kozłowski E, Antosz K, Parczewski R (2023) A new approach to production process capability assessment for non-normal data. Appl Sci 13(11):6721. https://doi.org/10.3390/app13116721

Denysenko, Y., Ivanov, V., Luscinski, S., Zaloga, V.: An integrated approach for improving tool provisioning efficiency. Management and Production Engineering Review 11(4), pp. 4–12 (2020). https://doi.org/10.24425/mper.2020.136115

Gasanov M, Kotliar A, Basova Y, Ivanova M, Panamariova O (2019) Increasing of lathe equipment efficiency by application of gang-tool holder. In: Gapiński B, Szostak M, Ivanov V (eds) Advances in manufacturing II. MANUFACTURING 2019. Lecture notes in mechanical engineering, Springer, Cham, pp 133–144. https://doi.org/10.1007/978-3-030-16943-5_12

Grześkowiak M, Trojanowska J (2021) Production line balancing in a mixed-model production system: a case study. In: Tonkonogyi V et al (eds) Advanced manufacturing processes II. InterPartner 2020. Lecture notes in mechanical engineering, Springer, Cham, pp 24–32. https://doi.org/10.1007/978-3-030-68014-5_3

Jasiulewicz-Kaczmarek M, Antosz K, Waszkowski R, Nielsen I, Čep R (2023) "Technology" as the fourth dimension of sustainable maintenance management. IFAC-PapersOnLine 56(2):162–167. https://doi.org/10.1016/j.ifacol.2023.10.1563

Jasiulewicz-Kaczmarek M, Antosz K, Zhang C, Waszkowski R (2022) Assessing the barriers to industry 4.0 implementation from a maintenance management perspective—pilot study results. IFAC-PapersOnLine 55(2):223–228. https://doi.org/10.1016/j.ifacol.2022.04.197

Karpus VE, Ivanov VA (2012) Choice of the optimal configuration of modular reusable fixtures. Russ Eng Res 32:213–219. https://doi.org/10.3103/S1068798X12030124

Karpus' VE, Ivanov VA (2008) Universal-composite adjustable machine-tool attachments. Russian Eng Res 28:1077–1083. https://doi.org/10.3103/S1068798X08110105

Krol O, Sokolov V, Tsankov P (2023) Improvement of a two-stage drive for multioperational milling machine. In: Ivanov V, Trojanowska J, Pavlenko I, Rauch E, Piteľ J (eds) Advances in design, simulation and manufacturing VI. DSMIE 2023. Lecture notes in mechanical engineering, Springer, Cham, pp 264–273. https://doi.org/10.1007/978-3-031-32767-4_25

Krol O, Sokolov V, Logunov O (2024) Development of machining center rotary table with a modified worm drive. In: Machado J et al (2024) Innovations in mechanical engineering III. Icieng 2024. Lecture notes in mechanical engineering, Springer, Cham, pp 372–383. https://doi.org/10.1007/978-3-031-62684-5_33

Kushnirov P, Gelatko M, Dehtiarov I, Evtuhov A, Stupin B, Neshta A, Ostapenko B (2023) Milling heads for machining mutually perpendicular flat surfaces. MM Sci J 6441–6445. https://doi.org/10.17973/MMSJ.2023_06_2023008

Kusyi Y, Stupnytskyy V (2020) Optimization of the technological process based on analysis of technological damageability of castings. In: Ivanov V, Trojanowska J, Pavlenko I, Zajac J, Peraković D (eds) Advances in design, simulation and manufacturing III. DSMIE 2020. Lecture notes in mechanical engineering, Springer, Cham, pp 276–284. https://doi.org/10.1007/978-3-030-50794-7_27

Kusyi Y, Stupnytskyy V, Onysko O, Dragašius E, Baskutis S, Chatys R (2022) Optimization synthesis of technological parameters during manufacturing of the parts. Eksploatacja i Niezawodność—Mainten Reliabil 24(4):655–667. https://doi.org/10.17531/ein.2022.4.6

Kusyi Y, Kusen N, Slipchuk A, Schuliar I Pituley L (2024) Design of the technological route for sustainable machining of functional surfaces for automotive engineering parts. In: Ivanov V, Trojanowska J, Pavlenko I, Rauch E, Piteľ J (eds) Advances in design, simulation and manufacturing VII. DSMIE 2024. Lecture notes in mechanical engineering, Springer, Cham, pp 263–275. https://doi.org/10.1007/978-3-031-61797-3_22

Muzylyov D, Shramenko N, Karnaukh M (2021) Choice of carrier behavior strategy according to industry 4.0. In: Ivanov V, Trojanowska J, Pavlenko I, Zajac J, Peraković D (eds) Advances in design, simulation and manufacturing IV. DSMIE 2021. Lecture notes in mechanical engineering, Springer, Cham, pp 213–222. https://doi.org/10.1007/978-3-030-77719-7_22

Shevchenko S, Mukhovaty A, Krol O (2020) Modification of two-stage coaxial gearbox. In: Radionov AA, Gasiyarov VR (eds) Proceedings of the 6th international conference on industrial engineering (ICIE 2020). ICIE 2021. Lecture notes in mechanical engineering. Springer, Cham, pp 28–35. https://doi.org/10.1007/978-3-030-54814-8_4

Shramenko N, Muzylyov D (2020) Forecasting of overloading volumes in transport systems based on the fuzzy-neural model. In: Ivanov V et al (eds) Advances in design, simulation and manufacturing II. DSMIE 2019. Lecture notes in mechanical engineering, Springer, Cham, pp 311–320. https://doi.org/10.1007/978-3-030-22365-6_31

Sokolov V, Porkuian O, Krol O, Baturin Y (2020) Design calculation of electrohydraulic servo drive for technological equipment. In: Ivanov V, Trojanowska J, Pavlenko I, Zajac J, Peraković D (eds) Advances in design, simulation and manufacturing III. DSMIE 2020. Lecture notes in mechanical engineering, Springer, Cham, pp 75–84. https://doi.org/10.1007/978-3-030-50794-7_8

Szczepaniak M, Trojanowska J (2019) Methodology of manufacturing process analysis. In: Hamrol A, Kujawińska A, Barraza M (eds) Advances in manufacturing II. MANUFACTURING 2019. Lecture notes in mechanical engineering, Springer, Cham, pp 281–294. https://doi.org/10.1007/978-3-030-18789-7_24

Tigariev V, Tonkonogyi V, Salii V, Klimenko S, Dovhan A (2020) Designing in modern CAD using information model. In: Tonkonogyi V et al (eds) Advanced manufacturing processes. Inter-Partner 2019. Lecture notes in mechanical engineering, Springer, Cham, pp 331–341. https://doi.org/10.1007/978-3-030-40724-7_34

Varela MLR, Putnik GD, Manupati VK et al (2021) Integrated process planning and scheduling in networked manufacturing systems for I4.0: a review and framework proposal. Wireless Netw 27:1587–1599. https://doi.org/10.1007/s11276-019-02082-8

Yakovenko I, Permyakov A, Ivanova M, Basova Y, Shepeliev D (2021) Lifecycle management of modular machine tools. In: Tonkonogyi V, Ivanov V, Trojanowska J, Oborskyi G, Pavlenko I (eds) Advanced manufacturing processes III. InterPartner 2021. Lecture notes in mechanical engineering, Springer, Cham, pp 127–137. https://doi.org/10.1007/978-3-030-91327-4_13

Yakovenko I, Permyakov A, Dobrotvorskiy S, Basova Y, Kotliar A, Zinchenko A (2023) Prospects for the development of process equipment in aggregate-modular design for sustainable mechanical engineering. Int J Mechatron Appl Mech 13:145–156. https://doi.org/10.17683/ijomam/issue13.18

Zakharkin O (2004) Technological fundamentals of manufacturing engineering. Sumy State University, Sumy. ISBN 966-657-019-X [in Ukrainian]

Zaleta OM, Povstyanoy OY, Ribeiro LF, Redko RG, Bozhko TY, Chetverzhuk TI (2023) Automation of optimization synthesis for modular technological equipment. J Eng Sci 10(1):A6-A14. https://doi.org/10.21272/jes.2023.10(1).a2

Chapter 3
Locating

3.1 Fundamental Principles

When assembling machine elements, it is necessary to ensure the correct positioning of parts and units in the assemblies. When machining workpieces, they must be correctly oriented to the elements of the technological system. The relative orientation of products in assemblies and workpieces during machining is achieved by locating.

Locating sets a workpiece or product to a desired position relative to the selected coordinate system (ISO 5459 2011). Workpiece locating during machining is provided by fixtures of various designs or by marking workpieces before machining, i.e., by drawing lines, dots, etc., on their surfaces. The position achieved during locating is fixed when clamping the workpiece. Workpiece positioning before machining demands the completion of two tasks: locating and clamping.

It is known that any material body in three-dimensional space has six degrees of freedom (three movements along coordinate axes and three rotations around these axes). When a body is located, a certain number of positional constraints (movement and rotation limiters) are imposed on it, depriving it of certain degrees of freedom. In general, the basic terms and definitions of locating are given in ISO 5459 (2011), Zakharkin (2004).

A datum is a surface or a combination of surfaces, an axis, or a point that belongs to the workpiece or product and is used for locating. A contact point is a point that represents a relation between a workpiece and the selected coordinate system.

A locating chart is a layout of the contact points on workpiece datums. All contact points are marked with a numeric icon, starting with the datum that has the most significant number of contact points. In any projection, one point is displayed by superimposing one contact point on another, and two numbers are placed next to it. In this way, the numerical value of the linear or angular position of the product at the corresponding coordinate is determined.

In actual locating, positional constraints are replaced by contact between the corresponding surfaces or contact points of products during assembly or between

V. Ivanov et al., *Fundamentals of Manufacturing Engineering Using Digital Visualization*, Springer Tracts in Mechanical Engineering,
https://doi.org/10.1007/978-3-031-74360-3_3

a workpiece and a fixture. The number of contact points must equal the number of positional constraints they replace. Upon that, the contact point deprives the product of one degree of freedom.

There are conventional symbols (Table 3.1) for marking contact points. Design documentation is analyzed using symbols mentioned in the study (Zakharkin 2004).

Contact points are indicated as follows: side view (¬V‾) and top view (-◇-).

Manufacturing engineers use (Zakharkin 2004) to indicate process datums on a locating chart when machining products. The manufacturing engineer indicates the marking of a support, the shape of a contact surface, and the number of degrees of freedom the product is deprived of. Some of the most used are listed in Table 3.1.

3.2 Classification

The entire variety of parts surfaces is classified into three types (functional, reference, and free).

Functional surfaces directly serve the purpose of the part (pulley surface in contact with the drive belt, cutting surface in screw mechanisms, working surface of wheel teeth, blade surfaces interacting with the operating environment in solid, liquid, gaseous states, reflector surfaces of light, heat, and other flows).

Reference surfaces determine the position of the part in a higher-level product or the position of other products attached to it.

Free surfaces do not come into contact with surfaces of other products but determine dimensions, weight, stiffness, and other parameters of parts. These surfaces may or may not be processed, i.e., remain in the state of the original workpiece.

All these surfaces comprise a set of connected and unconnected surfaces. The former perform specific functions as functional or reference surfaces, while the latter are free.

Datums are classified according to ISO 5459 (2011) in purpose, number of degrees of freedom, and exposure character (Fig. 3.1).

Datums are systemized according to purpose into design datums (primary and supplementary), process datums, and measuring datums.

According to ISO 5459 (2011), datums are generally classified by the number of freedom degrees in descending order. In other words, if one of the datums deprives a workpiece of three degrees of freedom, and the other two datums deprive it of two and one degree of freedom, respectively, the first datum is called the principal or primary datum. Then, the datum depriving the workpiece of two degrees of freedom is called the secondary datum, and the datum depriving the workpiece of one degree of freedom becomes the tertiary datum, respectively. This system creates a locating hierarchical structure (Fig. 3.2).

However, from an information viewpoint, this is inconvenient, so (Zakharkin 2004; Yakovenko et al. 2022) suggests naming datums by their functional purpose (Table 3.2), i.e., the name already includes both the number of degrees of freedom and the directions in which these degrees of freedom are maintained. That's easier to

Table 3.1 Supports and their designation on locating charts

Type of support/shape of a contact surface /locating element	Designation of a support/shape of a contact surface/locating element in view		
	side	top	bottom
Supports			
Fixed support	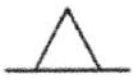		
Moving support	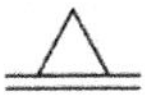	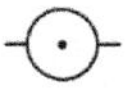	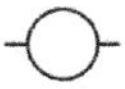
Tilting support		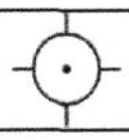	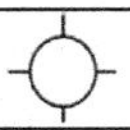
Shape of a contact surface			
Plane		–	–
Spherical		–	–
Cylindrical		–	–
V-shaped		–	–
Rhombic		–	–
Conical		–	–
Locating element			
Solid center		–	–
Live center		–	–
Tilted center		–	–
Three-jaw chuck		–	–
Centre chuck		–	–
Follow rest		–	–
Steady rest		–	–

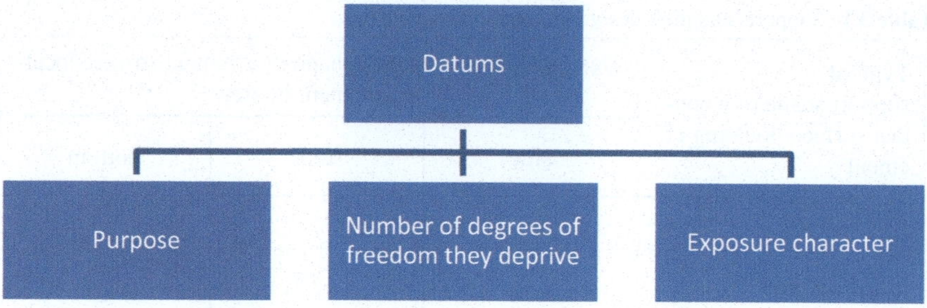

Fig. 3.1 Classification of datums

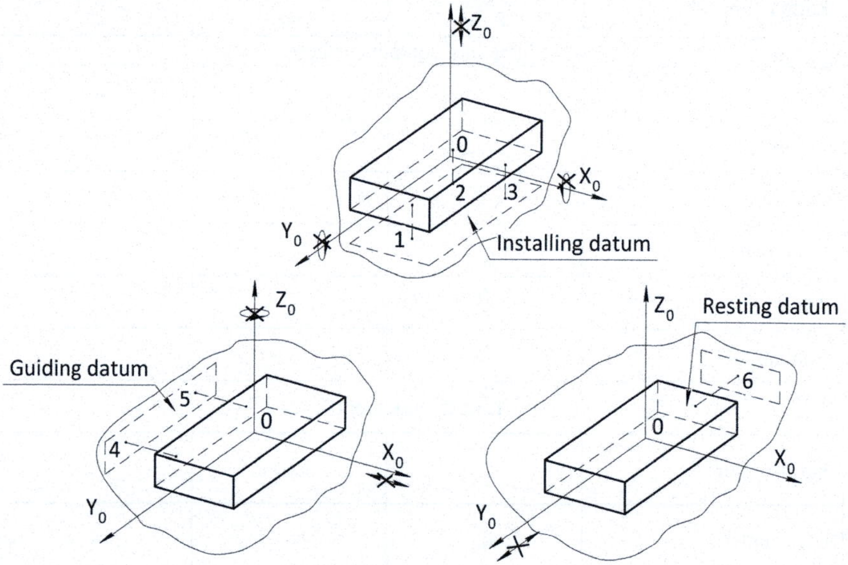

Fig. 3.2 Installing, guiding, and resting datums

understand and can immediately indicate how a particular workpiece is fitted, even without an operational sketch.

The datums, classified by the character of exposure, can be real and latent (Fig. 3.3). A real datum is a datum of a workpiece or product in the form of an actual surface, a marking stroke, or a point of strokes‘ intersection. A latent datum is a datum of a workpiece or product in the form of an imaginary plane, axis, or point (symmetry plane, axis, point).

All workpiece surfaces are generally divided into two groups:

- not processed after a workpiece has been manufactured;
- processed with a given accuracy.

Table 3.2 Datums' characteristics according to the number of freedom degrees they deprive

Datum	Degree of freedom deprived	Surface type employing the datum
Installing	3 (one movement, two rotation)	Plane
Guiding	2 (one movement, one rotation)	Plane
Resting	1 (one movement or one rotation)	Plane, diamond pin
Double-guiding	4 (two movement, two rotation)	Long cylinder
Double-resting	2 (two movement)	Short cylinder

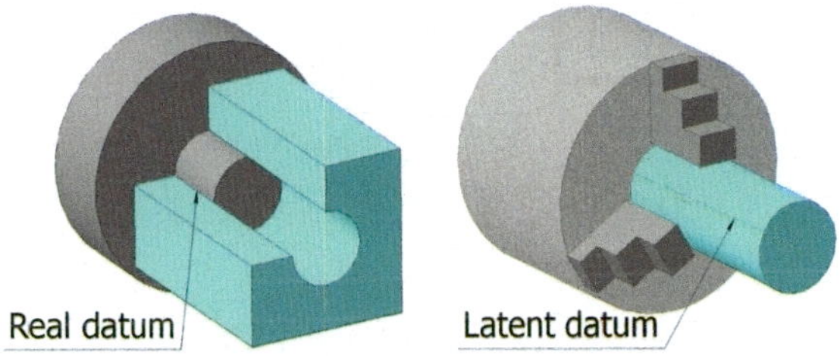

Fig. 3.3 Double-resting datum (real and latent)

The method of manufacturing a workpiece ensures the accuracy and relative positioning of the surfaces of the first group. Selecting the proper process datum ensures the accuracy of the surface's relative positioning of the first and second groups.

Process datums are divided into rough and finishing datums. There are specific rules for their assignment.

Rough datums are those used in the first machining operation. They are intended to prepare the finishing data and are used only once. The following principles should be considered when selecting them:

- surfaces that cannot be machined should be used as rough datums;
- die parting surfaces, those with gates, and other similar features of workpieces should not be used as rough datums;
- surfaces with a minimum allowance should be used as rough datums.

Finishing datums are used for finishing machining. The following principles are essential to consider when choosing them:

- finishing datums should be elements of the part that is its primary design or measuring datums (the principle of combining datums);
- finishing datums should ensure the machining of workpieces in various technological operations without changing the datums (the principle of constancy of datums);
- if a workpiece lacks such surfaces, artificial finishing datums are created (e.g., center holes, artificial sprues, etc.).

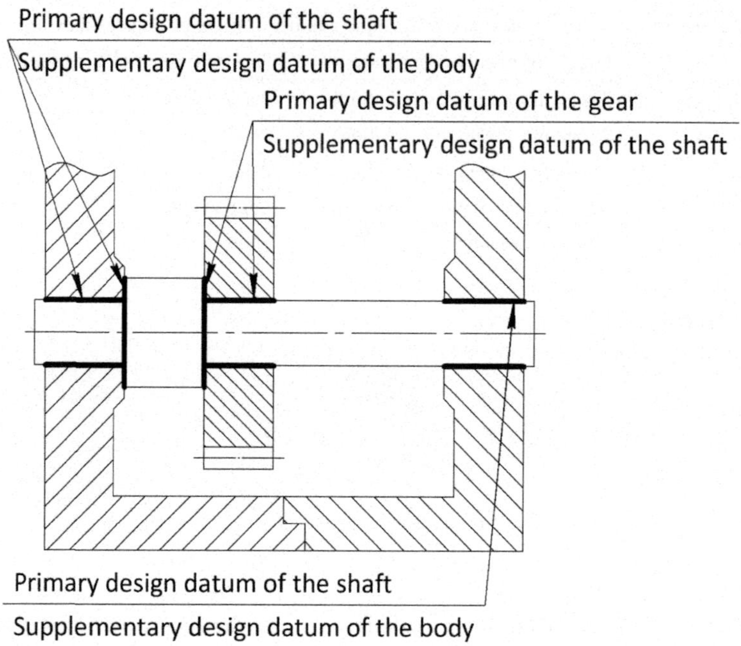

Fig. 3.4 Primary and supplementary design datums

3.3 Design Datums

A design datum determines the position of a part or assembly unit in a product. In turn, design datums are divided into primary and supplementary datums.

A primary design datum belongs to a given part and determines its position in a product (Fig. 3.4). A supplementary design datum belongs to a given part or assembly unit and determines the position of parts and assemblies attached to it (Fig. 3.4).

3.4 Process Datums

A process datum determines the position of a workpiece or product during a manufacturing or repair process (Fig. 3.5). Process datums are generally divided into the following:

- process datums for assembly—a surface, line, or point of a part or assembly unit to which other parts or assembly units of a product are referenced;
- process datums for machining—a surface, line, or point of a workpiece relative to which the machined surfaces are oriented on a given setup.

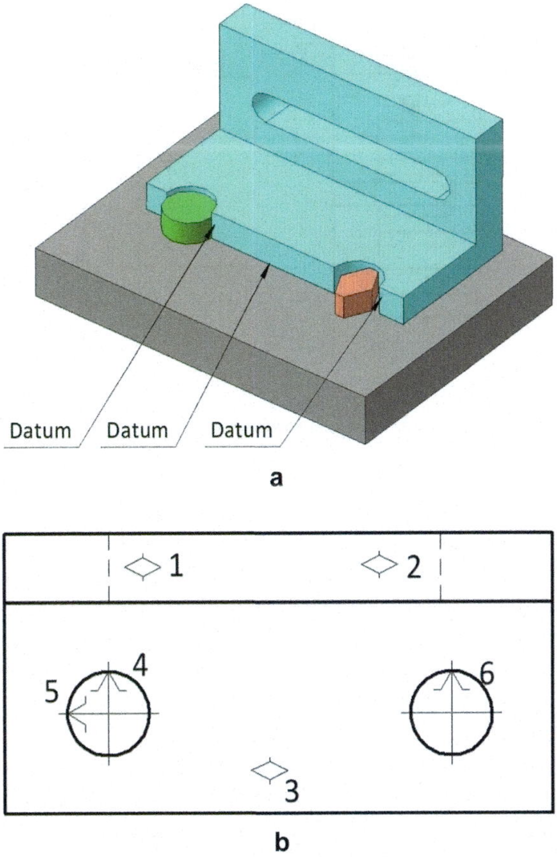

Fig. 3.5 Process datums: **a**—3D model; **b**—locating chart

Marking lines and dots on the physical surfaces of workpieces are also used as process datums to align them with fixtures that determine the path of cutting tools.

Process datums can be real (actual product surfaces) and artificial (specifically designed for manufacturing, assembly, or repair, such as center holes on a shaft).

According to the specifics of their application, process datums used in machining are divided into contact, adjustment, and verification datums.

Contact datums are process datums that come into direct contact with the corresponding locating elements of fixtures. Contact process datums are widely used when working on well-coordinated machine tools in large-batch manufacturing.

Adjustment datums are workpiece surfaces to which machined surfaces are oriented and which are directly related to them by dimensions. They are created by the same setup as the workpiece surfaces under consideration.

The adjustment datum is directly dimensionally related to a workpiece resting datum. The latter is a process datum for obtaining linear dimensions only when machining the adjustment datum, which is directly dimensionally related (Fig. 3.6).

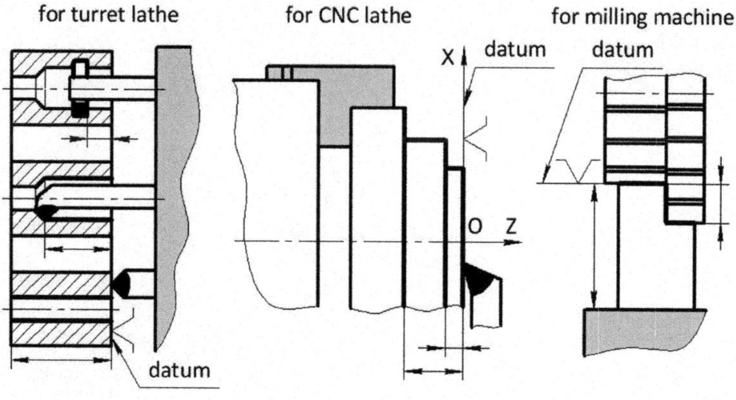

Fig. 3.6 Example of adjustment datums

The adjustment datum is the process datum for machining all other surfaces. Depending on the configuration and requirements for a workpiece, it can have several adjustment datums in the same dimensional direction. This condition complicates machine adjustment to some extent but allows direct dimensioning between surfaces whose relative position is important for the final product.

Verification process datums include a surface, line, or point of a workpiece or part against which the position of a workpiece on a machine tool or the positioning of a cutting tool during workpiece machining is verified, as well as the position of other parts or assembly units during product assembly (Fig. 3.7).

This method does not require complex fixtures to orient a workpiece on a machine tool, unlike the machining method using resting datums. Practice shows that the verification process datum used in assembly and machining can be material (imaginary) or conditional (latent).

Fig. 3.7 Example of verification datums

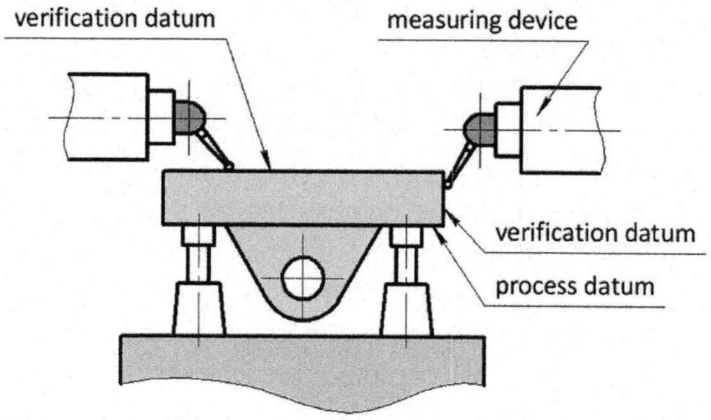

3.5 Measuring Datums

A measuring datum is a datum of a part or assembly unit used to measure dimensions to be machined or assembled or to verify the relative position of parts' surfaces or product elements (Fig. 3.8).

The measuring datum in a design is related to control surfaces of the part by direct dimensions or conditions (Pop et al. 2019). Usually, the measuring datum coincides with a design datum. If the measuring datum is a tangible surface, a measurement is made by conventional direct measurement methods. Suppose the measuring datum is an immaterial element (angle bisector, axial line, plane of symmetry, etc.). It is materialized using auxiliary parts (e.g., pins, rollers, tensioned strings, etc.), optical installations (collimators), and other devices.

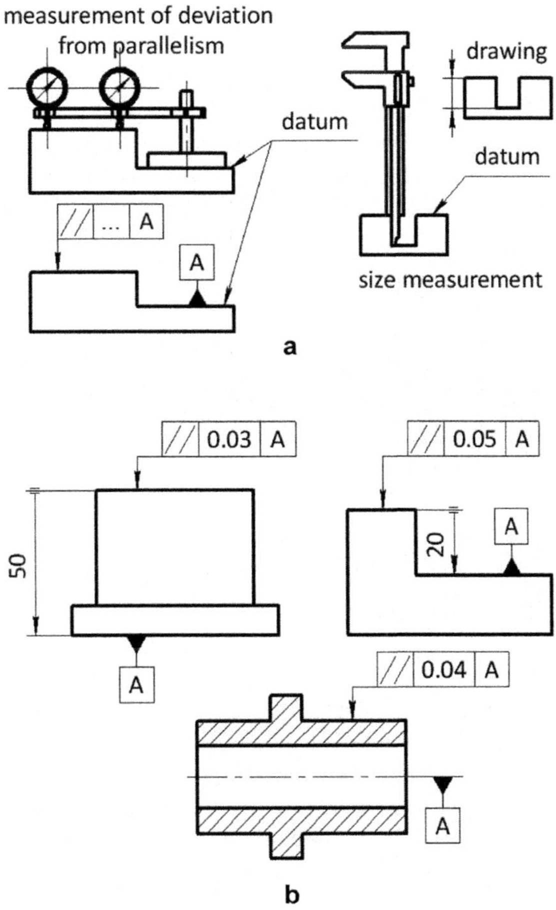

Fig. 3.8 Measuring datums: **a**—size control; **b**—drawing designation

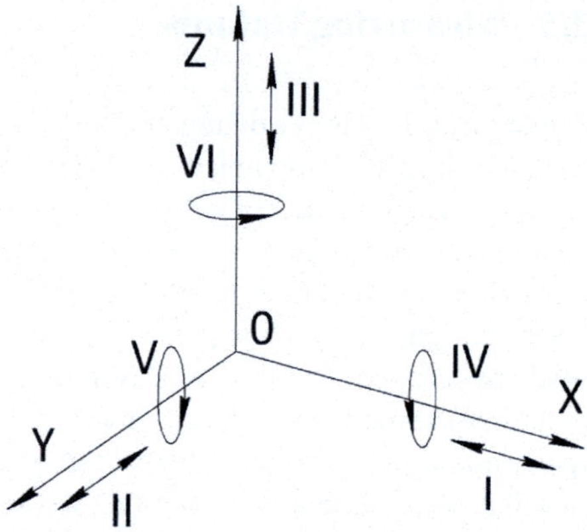

Fig. 3.9 Coordinate system and degrees of freedom

3.6 Theoretical Locating Charts

To completely locate the product in space, it must be deprived of six degrees of freedom (Virasak 2019). It is considered a three-dimensional coordinate system with axes X, Y, and Z centered on O (Fig. 3.9). The degrees of freedom are divided into linear and angular coordinates. Linear coordinates reflect movement along the X, Y, and Z-axes and correspond to degrees of freedom I, II, and III. Angular coordinates reflect rotation around the X, Y, and Z-axes and correspond to degrees of freedom IV, V, and VI.

SCAN QR ManuVis Manufacturing Visualization

To access the digital content in this subchapter, please follow these steps:
(1) open the ManuVis application on your device;
(2) scan the QR code

Further material describes theoretical locating charts for workpieces of different classes. A theoretical chart simulates locating workpieces such as housings, plates, and similar parts. Theoretical mechanics suggest that determining the position of a prismatic body relative to the OXYZ coordinate system demands connecting its lower surface A with three rigid bilateral coordinate connections to the plane XOY of the rectangular coordinate system (Fig. 3.10).

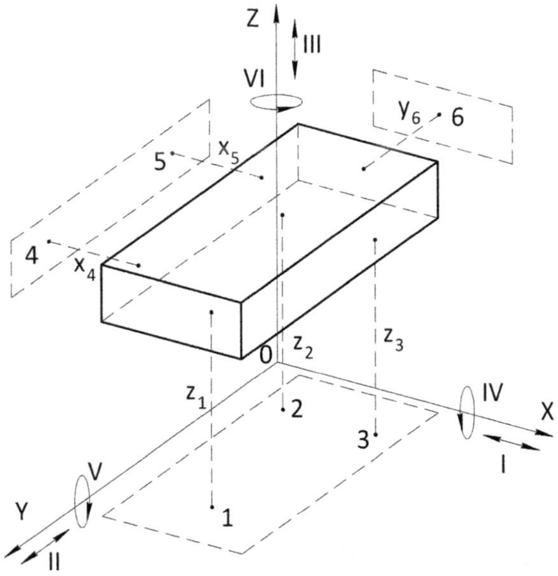

Fig. 3.10 Theoretical locating chart for the prismatic parts

The z-links prevent the body from moving along the OZ coordinate axis and rotating around the OY and OX axes. Considering surface, bearing three contact points and depriving the body of 3 degrees of freedom (movement along one of the coordinate axes and rotation around the other two coordinate axes) are a primary datum. It is also called an installing datum. The body's equilibrium under gravity determines the location of contact points.

Surface B has two coordinate connections, X_4 and X_5, to the plane ZOY, respectively, to prevent the body from moving along the OX axis and rotating about the OZ axis. Surface B deprives the body (workpiece) of two degrees of freedom (movement along one coordinate axis and rotation around the other), which is a secondary datum. It is also called a guiding datum. Thus, the body can only move along the OY axis. Surface C should be connected to the plane ZOX by a single rigid coordinate connection to prevent this movement. Surface C, bearing one datum and depriving the body of one degree of freedom (movement along one of the coordinate axes), is a tertiary datum. It is also called a resting datum.

The set of three datums forming the coordinate system of a workpiece (product) is a datum system. The theoretical locating chart is implemented by positioning the workpiece to locate fixture elements. Applying clamping forces ensures a non-separable contact of the datums with the locating elements.

The correspondences of coordinate connections, degrees of freedom, and datums they form are given in Table 3.3.

Table 3.3 Datums of the prismatic parts locating

Contact points	Degrees of freedom	Linear (L) and angular (α) coordinates	Coordinates			Locating datum
			X	Y	Z	
1, 2, 3	III, IV, V	L	0	0	1	Primary datum (installing datum)
		α	1	1	0	
4, 5	I, VI	L	1	0	0	Secondary datum (guiding datum)
		α	0	0	1	
6	II	L	0	1	0	Tertiary datum (resting datum)
		α	0	0	0	

SCAN QR ManuVis Manufacturing Visualization

To access the digital content in this subchapter, please follow these steps:
(1) open the ManuVis application on your device;
(2) scan the QR code

The following theoretical locating chart demonstrates how a workpiece is located when using long cylindrical surfaces, ends, and radial elements in fixtures with V-blocks.

By connecting the cylindrical surface A (Fig. 3.11) with two rigid bilateral coordinated connections to the XOY plane and two to the YOZ plane, a cylindrical body loses four degrees of freedom, as the Z-links prevent the body from moving along the OZ axis and rotating around the OX axis, the X coordinated connections prevent the body from moving along the OX axis and rotating around the OZ axis.

Surface A, bearing four datum points and depriving the body of four degrees of freedom (movement along two coordinate axes and rotation around these same axes), is called a primary datum. It is also called a double-guiding datum. Connecting the body's end C with a bilateral coordinate connection, the y coordinate with the XOZ plane, eliminates the possibility of moving the body along the OY axis. Surface C deprives the body of one degree of freedom. The movement along the OY axis is called a secondary datum. It is also called a resting datum. To deprive the body of the sixth degree of freedom (the ability to rotate around its axis), a sixth bilateral coordinate connection must be provided in the form of a datum located on the surface of the keyway flute B. Surface B, bearing one datum and depriving the body of one degree of freedom (rotation around one of the coordinate axes), is called the tertiary datum. It is also called a resting datum. The theoretical locating chart is most often implemented using V-blocks.

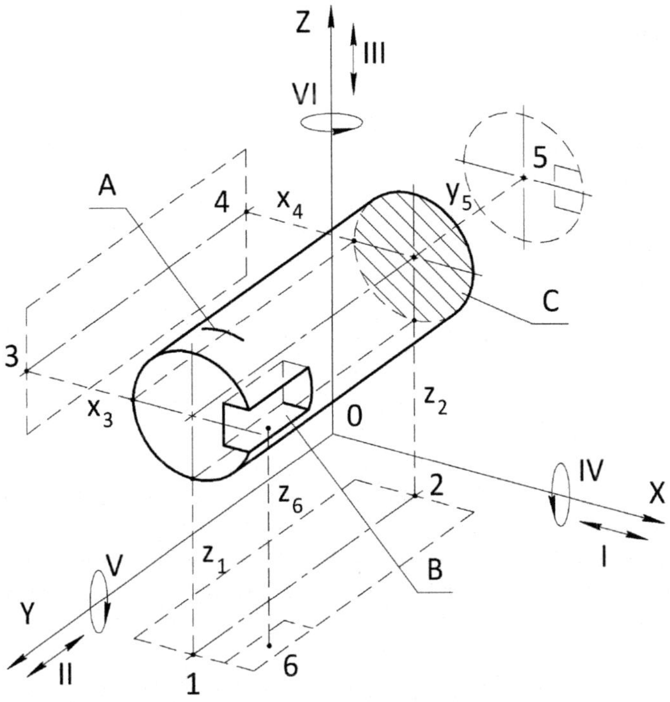

Fig. 3.11 Theoretical locating chart for long cylinders

The correspondences of coordinate connections, degrees of freedom, and datums they form are given in Table 3.4.

Table 3.4 Datums of the long cylinders' locating

Contact points	Degrees of freedom	Linear (L) and angular (α) coordinates	Coordinates			Locating datum
			X	Y	Z	
1, 2, 3, 4	I, III, IV, VI	L	1	0	1	Primary datum (double-guiding datum)
		α	1	0	1	
5	II	L	0	1	0	Secondary datum (resting datum)
		α	0	0	0	
6	V	L	0	0	0	Tertiary datum (resting datum)
		α	0	1	0	

To access the digital content in this subchapter, please follow these steps:
(1) open the ManuVis application on your device;
(2) scan the QR code

SCAN QR ManuVis Manufacturing Visualization

A disk is a cylindrical part where the length of a cylindrical surface is shorter than the diameter. As a result, orientation options for the short cylindrical surface are much more limited than for a long cylinder compared to end surfaces.

Accordingly, the orientation of a short cylindrical body (such as a thin disk) in space requires connecting its end surface C (Fig. 3.12) with three bilateral coordinate connections to the XOZ plane. In this case, the body loses three degrees of freedom (the ability to move along the OY axis and rotate around the OX and OZ axes).

Connecting its cylindrical surface B with bilateral coordinate connections is necessary to prevent the body from moving along the axes OX and OZ. In other words, the coordinates X_5 and Z_4 are aligned with the planes XOY and YOZ. The sixth bilateral coordinate connection, which prevents the body from rotating around its axis parallel to the OY axis, is created by placing a contact point on the surface of the keyway C.

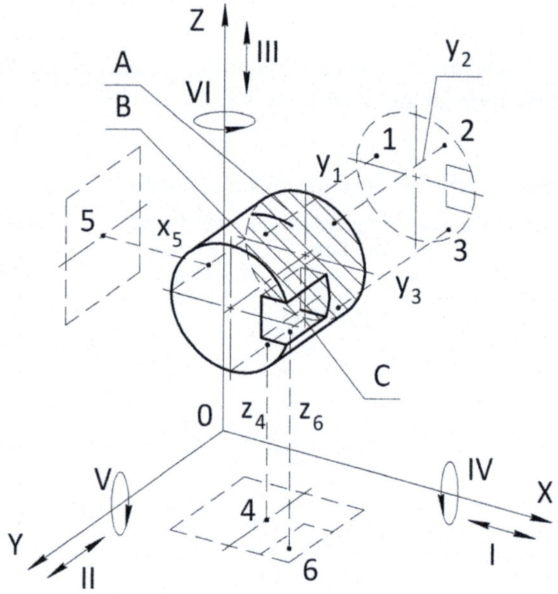

Fig. 3.12 Theoretical scheme of disks locating

With the appropriate replacement of bilateral coordinate connections with contact points, the end surface A of the disk, contacting three contact points and depriving the disk of three degrees of freedom, is called a primary datum. It is also called an installing datum. The cylindrical surface B, contacting two contact points and depriving the disk of two degrees of freedom, is called a secondary datum. It is also called a double-resting (or centering) datum. The surface of the keyway flute C, depriving the disk of one degree of freedom, is called a tertiary datum. It is also called a resting datum. Note that the charts for locating workpieces with internal cylindrical surfaces are fundamentally similar to those considered.

The correspondences of coordinate connections, degrees of freedom, and datums they form are given in Table 3.5.

To access the digital content in this subchapter, please follow these steps:
(1) open the ManuVis application on your device;
(2) scan the QR code

Notably, for a rotational part, depriving the workpiece of the sixth degree of freedom is only necessary if there are angular coordinate dimensions for surfaces to be machined in different operations. For example, depriving a workpiece of 5 degrees of freedom (partial locating) suffices to drill a hole in the workpiece (Fig. 3.13a).

Table 3.5 Datums of the disks locating

Contact points	Degrees of freedom	Linear (L) and angular (α) coordinates	Coordinates			Locating datum
			X	Y	Z	
1, 2, 3	II, IV, VI	L	0	1	0	Primary datum (installing datum)
		α	1	0	1	
4, 5	III, I	L	1	0	1	Secondary datum (double-resting datum)
		α	0	0	0	
6	V	L	0	0	0	Tertiary datum (resting datum)
		α	0	1	0	

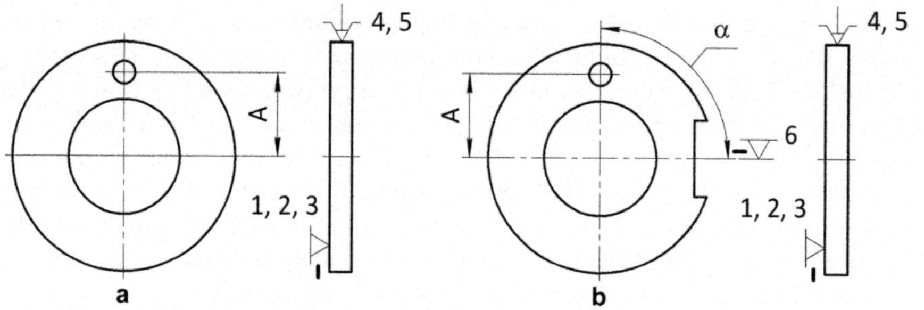

Fig. 3.13 Examples of incomplete (**a**) and complete (**b**) locating

The workpiece (Fig. 3.13b) must be additionally orientated relative to the previously machined radial slot and thus be deprived of another sixth degree of freedom (complete locating).

3.7 Case Studies

The most used datums for prismatic workpieces are planes (Fig. 3.14), planes and holes (Fig. 3.15), and other external surfaces of a workpiece (Fig. 3.16).

Shaft-type workpieces are typically located in centers (Figs. 3.17 and 3.18) or V-blocks (Fig. 3.19). Long shafts (with a ratio of $L/d > 12$) are machined using rests. Centers can be rigid, floating, or rotating. It is also possible to locate the shaft in a three-point chuck with a rear-rotating center tuck (Fig. 3.20). Some typical locating charts using these are provided below.

During machining, disk- and ring-type workpieces are mostly positioned in self-centering chucks (Fig. 3.21), on special mandrels (Fig. 3.22), and in other fixtures. Planes, cylindrical surfaces, and radial elements are used as datums. Some of the locating charts are demonstrated below.

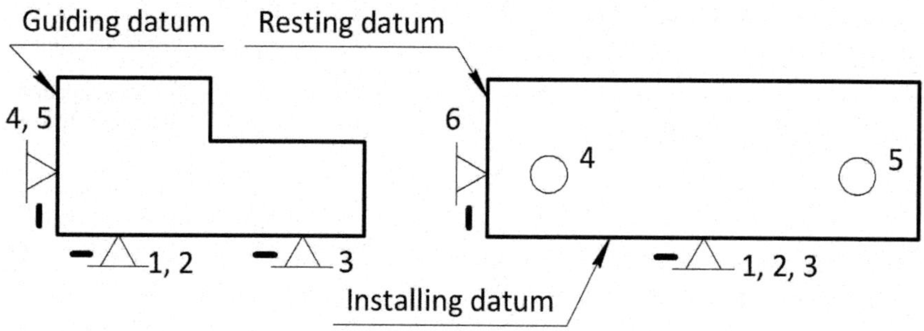

Fig. 3.14 Locating against three planes

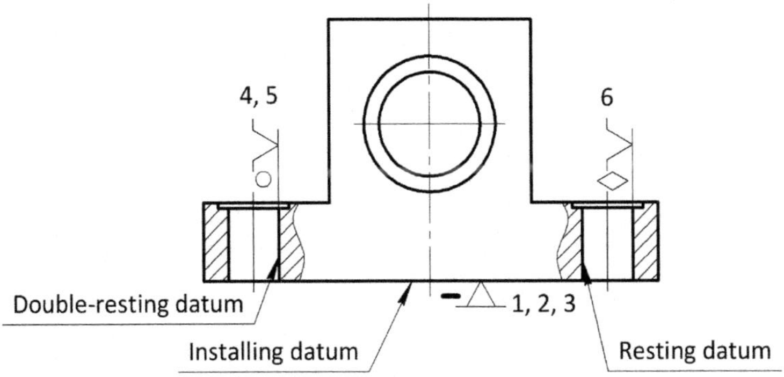

Fig. 3.15 Locating against one plane and two pins

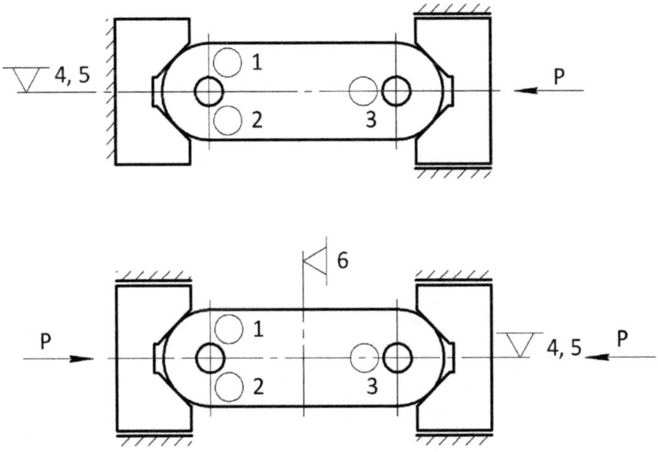

Fig. 3.16 Locating with V-blocks for a latent datum implementation

Fig. 3.17 Locating in left rigid and right rotating centers

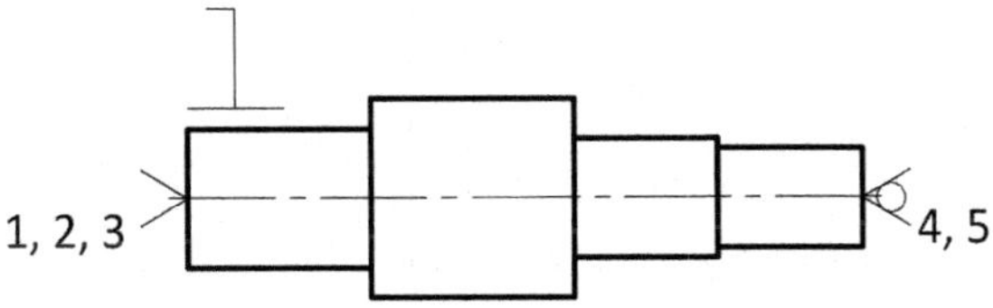

Fig. 3.18 Locating using left floating and right rotating centers

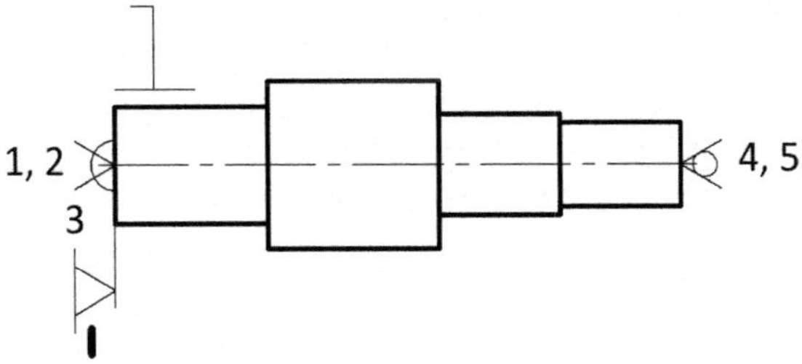

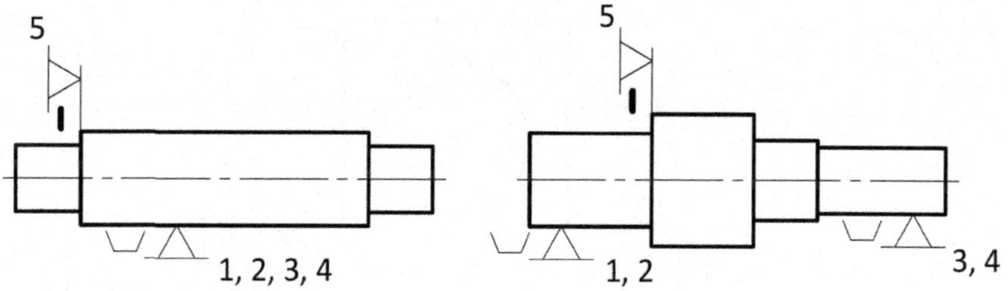

Fig. 3.19 Locating variants in V-blocks

Fig. 3.20 Locating in a chuck with a rear-rotating center

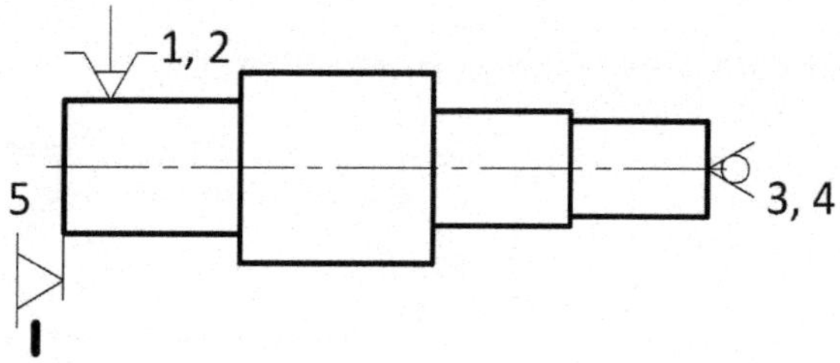

Fig. 3.21 Locating in a three-jaw chuck

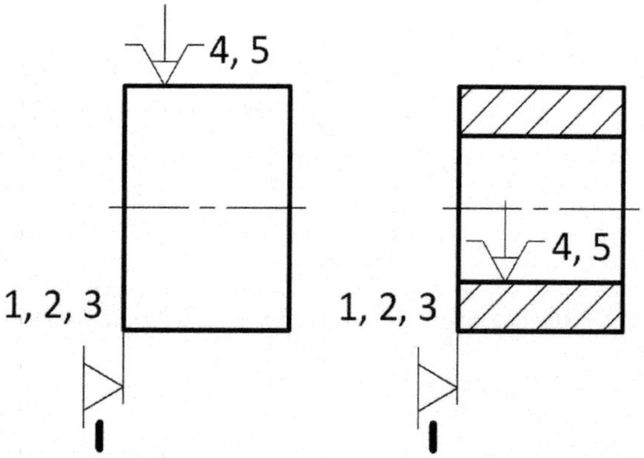

This is a partial list of possible locating cases for different workpieces. Some other cases are provided in the literature, and many other charts are derived from those discussed above.

Adjustment helps achieve the workpiece's required orientation relative to a cutting tool when performing automatic sizing on machine tools. A complete locating chart with the deprivation of six degrees of freedom is used if a surface is defined in all directions, i.e., three coordinate dimensions and three angular parameters are specified. If a smaller number of coordinates defines the surface to be machined, then there is no need to use complete locating. Instead, it is sufficient to deprive the body of

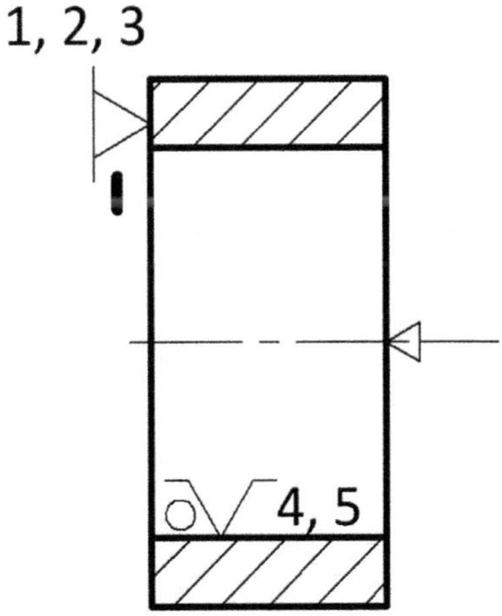

Fig. 3.22 Locating in a short mandrel

five, four, or three degrees of freedom. Such workpiece locating is called an incomplete locating. Thus, the fixture configuration would be easier and cheaper. Marks on a machine tool, inaccurate (or uncontrolled) stops, etc., can roughly orientate the workpiece in a direction not covered by the workpiece's locating.

Consider examples of locating a prismatic body with machining surfaces of different shapes (Fig. 3.23). Machining a semi-closed slot with an end-mill cutter, given the dimensions A_x, A_y, A_z, and surface directions relative to these axes, demands the total locating of the prismatic body, depriving it of all six degrees of freedom. The formula for locating is $3 + 2 + 1 = 6$.

Machining a through slot with a disk cutter (Fig. 3.24), given dimensions A_x, Az, and direction relative to the X, Y, and Z-axes, requires simplified locating, depriving a workpiece of five degrees of freedom. The formula for locating is $3 + 2 = 5$. It does not require the exact position of the workpiece on the Y-axis (i.e., along the

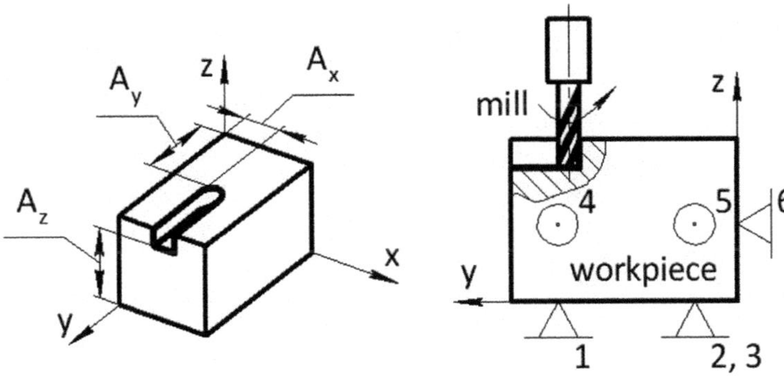

Fig. 3.23 Locating chart for machining of semi-closed slot

slot). The obligatory condition is that the workpiece is positioned in this direction within a tools' travel (L_{tt}).

Machining an upper plane of a part (Fig. 3.25), defined by the dimension A_Z, demands deprivation of three degrees of freedom: position along the Z-axis and rotation around the X and Y-axes. The formula for locating is $3 = 3$. The workpiece to be machined is located on three contact points. The machining zone must be limited by the width of a cylindrical cutter (B_{cc}) and the length of tool travel (L_{tt}), overlapping the machined surface. Flat grinding for workpieces located on a magnetic plate involves a similar approach.

Consecutive machining of a workpiece on a series of machine tools requires the exact positioning of supporting elements on all machines. When a workpiece is placed on three contact points, it must be positioned so the supporting triangle it forms is as large as possible (Fig. 3.26). The most preferred arrangement is one in which clamping, weight, and cutting forces are projected into the gravity center of

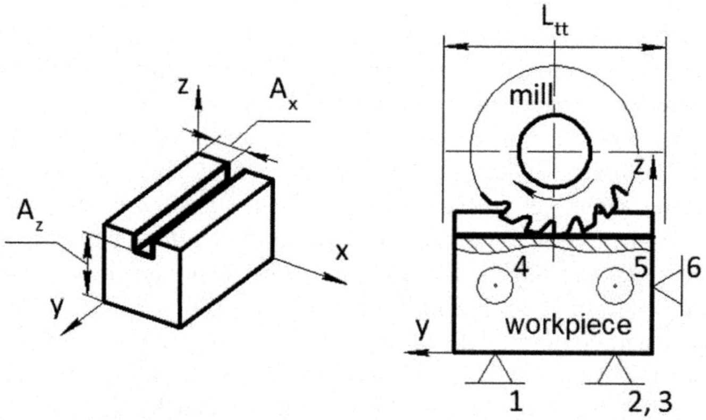

Fig. 3.24 Positioning and machining scheme of a closed slot

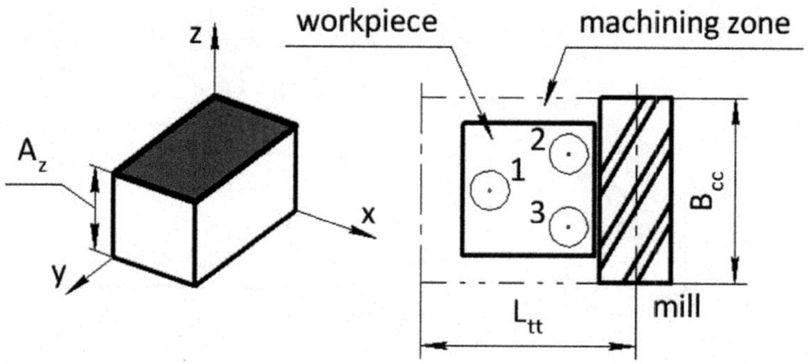

Fig. 3.25 Blank positioning scheme during surface machining

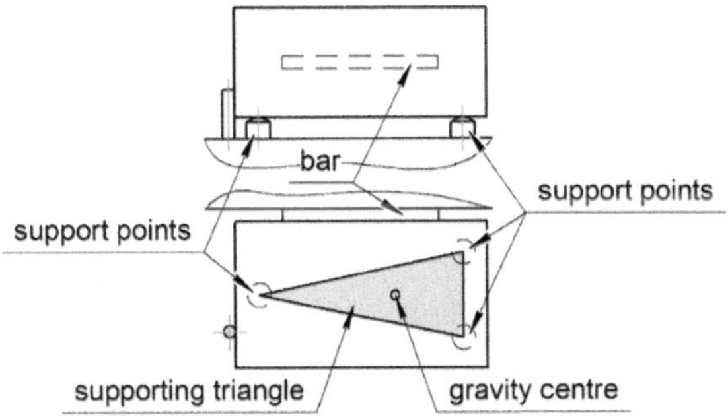

Fig. 3.26 Locating chart for a supporting triangle

the supporting triangle. Projection of the forces inside the supporting triangle is the least preferred in a normal machining process.

Locating a workpiece on a plate requires an installing datum to be larger than a fixture's installing surface and a guiding datum to be longer than a fixture's guiding surface. Contact areas of the primary (installing) and secondary (guiding) datums must be sufficient to prevent damage to the workpiece from excessive contact stress. This is particularly relevant for workpieces made of non-ferrous materials and alloys. When locating workpieces that have stamping or casting slopes, it is recommended to set installing surfaces of tools following the slope angles $\alpha = 5^{\circ} \div 7^{\circ}$.

The authors have contributed to fixture design knowledge (Karpus et al. 2019). Their expertise in locating has become crucial to the precise machining of different workpieces, particularly for shafts (Karpus and Ivanov 2012), connecting rods (Ivanov et al. 2022a, 2020a), lever-type parts (Ivanov et al. 2017, 2018), bracket-type parts (Ivanov et al. 2022b; Kolesnyk et al. 2024), and fork-type parts (Ivanov et al. 2021; Dehtiarov et al. 2024). The developed technical solutions are verified theoretically and experimentally, ensuring accuracy, stability, and repeatability in production. The previously obtained results have contributed to intelligent manufacturing, particularly in the decision-making of the locating charts (Ivanov et al. 2022c) and analysis of the technological features (Ivanov et al. 2020b). Fundamental results have been placed in implementing the process-oriented approach (Ivanov 2019) and developing a computer-aided fixture design system (Ivanov et al. 2016). This knowledge is crucial for optimizing manufacturing efficiency, reducing errors, and improving the quality of the final products.

References

Dehtiarov I, Ivanov V, Evtuhov A, Ostroha R, Ciszak O (2024) Research of static stiffness of flexible fixtures for y-type fork parts. In: Ivanov V, Trojanowska J, Pavlenko I, Rauch E, Pitel' J (eds) Advances in design, simulation and manufacturing VII. DSMIE 2024. Lecture notes in mechanical engineering, Springer, Cham, pp 164–177. https://doi.org/10.1007/978-3-031-61797-3_14

ISO 5459 (2011) (E): Geometrical product specifications (GPS)—geometrical tolerancing—datums and datum systems. 2nd edn. Switzerland, Geneva (2011). https://cdn.standards.iteh.ai/samples/40358/8913e16949684aa4889e9221ec7e8362/ISO-5459-2011.pdf

Ivanov V, Vashchenko S, Rong Y (2016) Information support of the computer-aided fixture design system. CEUR Workshop Proceed 1614:73–86

Ivanov V, Mital D, Karpus V et al (2017) Numerical simulation of the system "fixture–workpiece" for lever machining. Int J Adv Manuf Technol 91:79–90. https://doi.org/10.1007/s00170-016-9701-2

Ivanov V, Botko F, Dehtiarov I, Kočiško M, Evtuhov A, Pavlenko I, Trojanowska J (2022a) Development of flexible fixtures with incomplete locating: connecting rods machining case study. Machines 10:493. https://doi.org/10.3390/machines10070493

Ivanov V, Botko F, Kolos V, Pavlenko I, Hatala M, Antosz K, Trojanowska J (2022c) Locating chart choice based on the decision-making approach. Materials 15:3557. https://doi.org/10.3390/ma15103557

Ivanov V (2019) Process-oriented approach to fixture design. In: Ivanov V et al (eds) Advances in design, simulation and manufacturing. DSMIE 2019. Lecture notes in mechanical engineering, Springer, Cham, pp 42–50. https://doi.org/10.1007/978-3-319-93587-4_5

Ivanov V, Dehtiarov I, Denysenko Y, Malovana N, Martynova N (2018) Experimental diagnostic research of fixture. Diagnostyka 19(3):3–9. https://doi.org/10.29354/diag/92293

Ivanov V, Dehtiarov I, Zaloga V, Kosov I, Savchuk V (2020a) Increasing productivity of connecting rods machining. In: Ivanov V, Trojanowska J, Pavlenko I, Zajac J, Peraković D (eds) Advances in design, simulation and manufacturing III. DSMIE 2020. lecture notes in mechanical engineering, Springer, Cham, pp 264–275. https://doi.org/10.1007/978-3-030-50794-7_26

Ivanov V, Pavlenko I, Zaloga V, Liaposhchenko O, Pirogov D (2020) Technological features of locating charts in fixture design. In: Karabegović I (eds) New technologies, development and application III. NT 2020. Lecture notes in networks and systems 128, Springer, Cham, pp 66–74. https://doi.org/10.1007/978-3-030-46817-0_7

Ivanov V, Dehtiarov I, Evtuhov A, Pavlenko I, Ruban A (2021) Multiaxis machining of fork-type parts: fixture design and numerical simulation. In: Karabegović I (eds) New technologies, development and application IV. NT 2021. Lecture notes in networks and systems 233, Springer, Cham, pp 142–152. https://doi.org/10.1007/978-3-030-75275-0_17

Ivanov V, Kolos V, Liaposhchenko O, Pavlenko I (2022b) Technological assurance of bracket-type parts manufacturing. In: Knapčíková L, Peraković D, Behúnová A, Periša M (eds) 5th EAI international conference on management of manufacturing systems. EAI/Springer innovations in communication and computing, Springer, Cham, pp 397–409. https://doi.org/10.1007/978-3-030-67241-6_31

Karpus V, Ivanov V, Dehtiarov I, Zajac J, Kurochkina V (2019) Technological assurance of complex parts manufacturing. In: Ivanov V et al (eds) Advances in design, simulation and manufacturing. DSMIE 2019. Lecture notes in mechanical engineering, Springer, Cham, pp 51–61. https://doi.org/10.1007/978-3-319-93587-4_6

Karpus VE, Ivanov VA (2012) Locating accuracy of shafts in V-blocks. Russ Eng Res 32:144–150. https://doi.org/10.3103/S1068798X1202013X

Kolesnyk V, Kuruc M, Dehtiarov I, Neshta A, Avramenko S, Kononovych V, Dynnyk O, Dovhopolov A, Lemekh S, Titskyi R (2024) Experimental evaluation of flexible fixture stiffness for steering knuckles when loading a milling machine tool. Machines 12:405. https://doi.org/10.3390/machines12060405

Pop GM, Crisan LA, Tripa M, Neamtu C, Dragomir M (2019) On the association of datums and measurements using conventional measuring devices—topics within the GPS toolbox project. In: Majstorovic V, Durakbasa N (eds) Proceedings of the 12th international conference on measurement and quality control—cyber physical issue. IMEKOTC14 2019. Lecture notes in mechanical engineering, Springer, Cham, pp 209–216. https://doi.org/10.1007/978-3-030-18177-2_20

Virasak L (2019) Manufacturing processes 4–5. Open Oregon educational resources, Oregon. https://openoregon.pressbooks.pub/manufacturingprocesses45/
Yakovenko IE, Permyakov OA, Fesenko AV (2022) Technological basics of machine building. NTU «KhPI», Kharkiv
Zakharkin O (2004) Technological fundamentals of manufacturing engineering. Sumy State University, Sumy. ISBN 966-657-019-X [in Ukrainian]

Chapter 4
Product Quality

4.1 Accuracy Indicators

Manufacturing engineering considers accuracy to be the degree to which parameters of a finished product conform to the dimensions, shape, and other characteristics specified in a technical drawing.

A designer determines the level of manufacturing accuracy and surface roughness based on requirements to ensure the functional purpose of a product (Karpus and Ivanov 2012). Precision in manufacturing engineering is essential for improving the performance of machines and increasing their durability, reliability, speed, and other characteristics (Petrakov et al. 2024).

For example, (Yakovenko et al. 2022) found that reducing a gap in movable joints from 20 to 10 microns increases their service life from 740 to 1200 h.

The following indicators reveal the accuracy of machine parts' surfaces: dimensional accuracy, surface shape accuracy, accuracy of a relative positioning of surfaces, and surface roughness.

Accuracy assessment involves such notions:

- a basic size is a size determined by a designer when designing a product and rounded to a value from a normal range;
- an actual size is a size obtained as a result of processing and measured with the required accuracy;
- a quality class is a set of tolerances of the same accuracy level for a given dimensional range. There are 19 qualifications (01, 0, 1, 2, …, 17), with accuracy decreasing from quality class 01 to 17.

If not specified in the design, the shape accuracy of surfaces and their relative positioning accuracy may be within the tolerance for a particular dimension. If necessary, the designer may specify more precise requirements for the value of these quality parameters. They are normalized based on data (DSTU 2498–94 1994). Main shape

V. Ivanov et al., *Fundamentals of Manufacturing Engineering Using Digital Visualization*, Springer Tracts in Mechanical Engineering,
https://doi.org/10.1007/978-3-031-74360-3_4

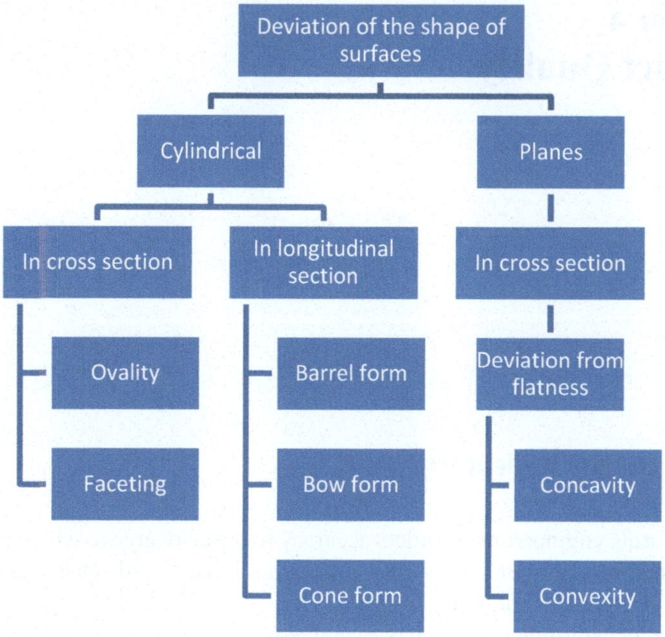

Fig. 4.1 Classification of some surface form deviations

deviations of cylindrical surfaces and planes are divided into the following types (Fig. 4.1).

Ovality (Fig. 4.2) is a deviation from roundness in which an actual profile is oval-shaped, with the largest and smallest diameters in mutually perpendicular directions. In some cases, an ovality tolerance may exceed a diameter tolerance, for example, when manufacturing thin-walled bushings that deform during machining and acquire the correct shape during assembly.

Faceting (Fig. 4.3) is a shape deviation from a circle, where an actual profile is multifaceted. Faceting means that the part's cross-sectional contour contains connected arcs of different radii. Faceting is classified according to several facets. In particular, faceting with an odd number of faces is characterized by the diameters of a cross-sectional profile being the same in all directions.

Barrel, bow, and cone shapes are considered in a longitudinal section as generatrices' deviations from parallelism (Figs. 4.4, 4.5, 4.6).

Errors in plane shapes are most often characterized by the following indicators: deviation from straightness and deviation from flatness. Deviations from flatness can be in the form of concavity or convexity (Figs. 4.7 and 4.8). The machining accuracy can be ensured by using the additional supports in fixture configurations, which provide higher stiffness (Kushnirov et al. 2024).

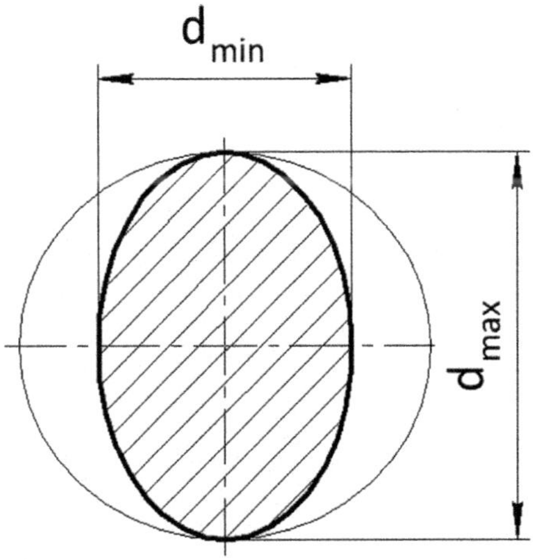

Fig. 4.2 Ovality

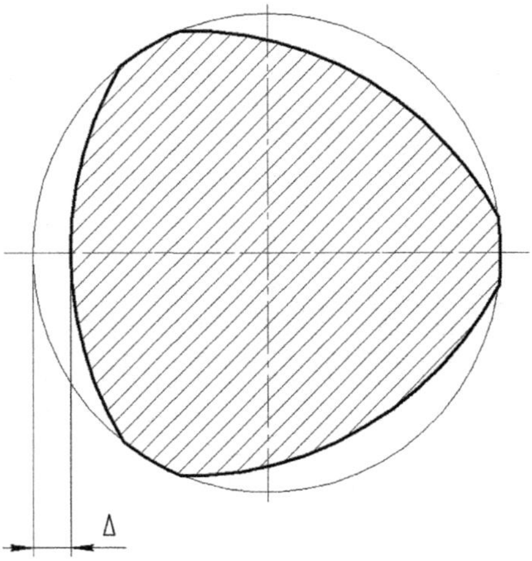

Fig. 4.3 Faceting

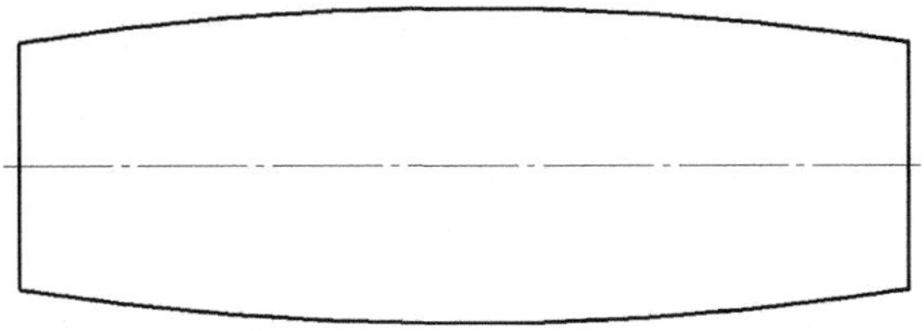
Fig. 4.4 Barrel form

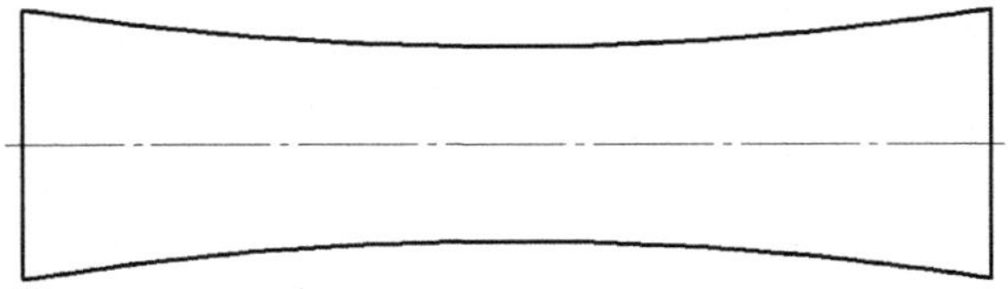
Fig. 4.5 Bow form

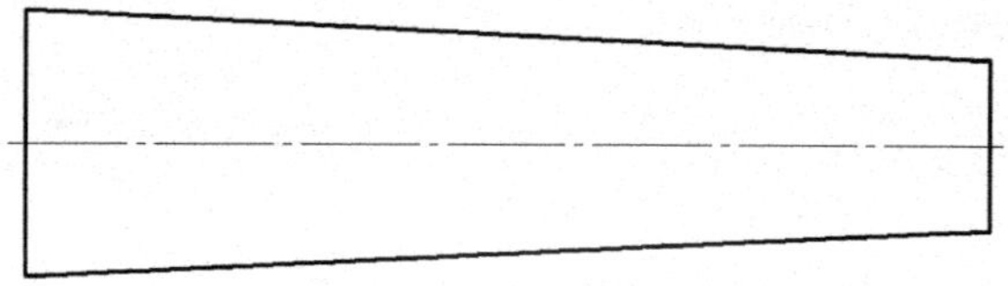

Fig. 4.6 Cone form

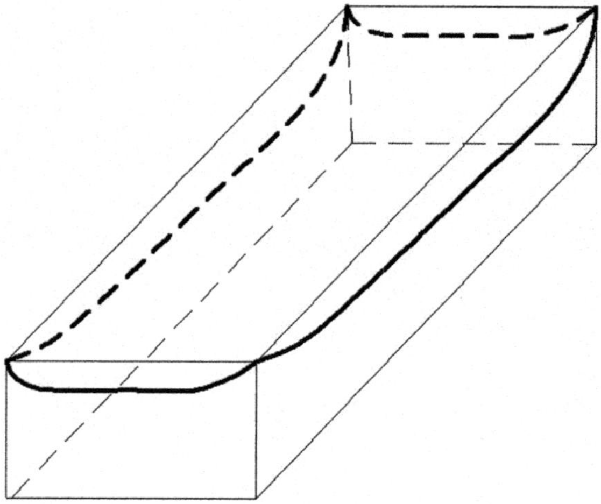

Fig. 4.7 Concavity

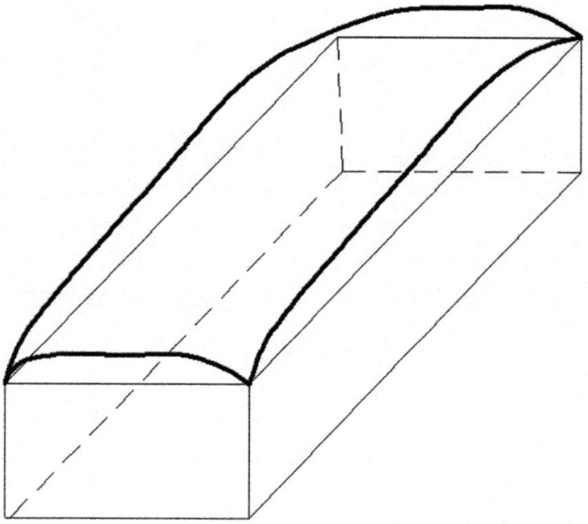

Fig. 4.8 Convexity

Deviation in a surface alignment is often assessed by such parameters as deviation from perpendicularity (Fig. 4.9a), parallelism (Fig. 4.9b), coaxiality (Fig. 4.9c), end runout, radial runout, etc.

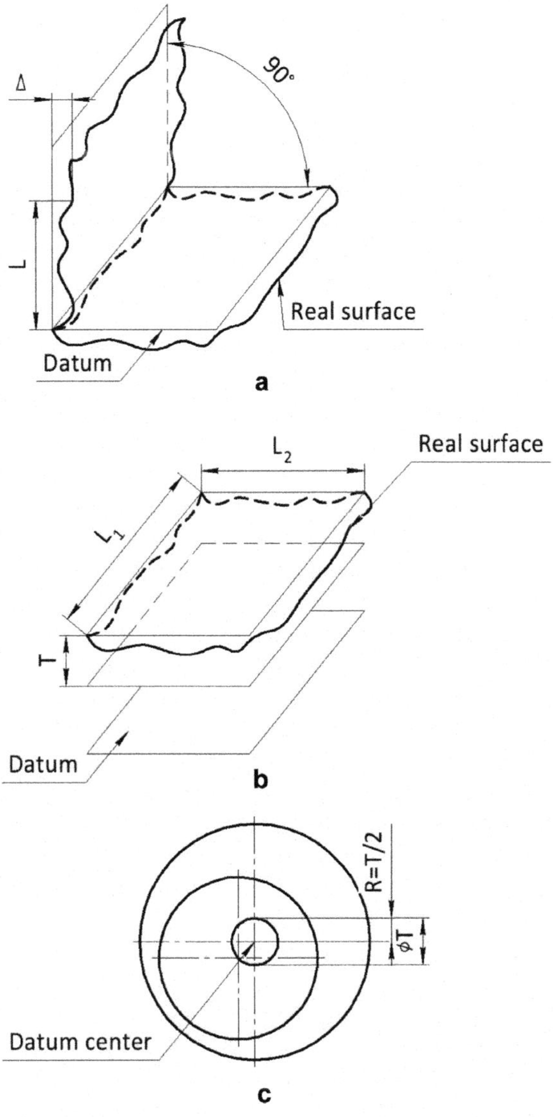

Fig. 4.9 Deviation types of surface alignment: **a**—perpendicularity; **b**—parallelism; **c**—coaxiality

Factors' analysis, causing such errors in machining and preventive measures development, is among the principal tasks of a process manager.

Digitalization is an important stage for manufacturing engineering and learning gain, particularly for accuracy indicators. The development and impact of various ICT tools and mobile applications designed to enhance engineering education (Pop et al. 2019), particularly in understanding geometrical tolerances (Pop et al. 2022) and form deviations (Pop et al. 2021), are examined in these papers.

4.2 Machining Accuracy

There are two (fundamentally different) methods of achieving dimensional accuracy in machining: a trial runs and measuring method and a method of automatic accuracy on a pre-adjusted machine.

The trial runs and measuring method involve bringing a tool to a workpiece from above and removing chips from a small area of the workpiece. After that, the machine is stopped, a test measurement of the resulting size is made, a deviation from the required size is determined, and the tool position is corrected. After that, the trial run and measuring are performed again. Once the required size is achieved, the entire surface is machined. This machining method is used in small batches, single pieces, and experimental production.

The method's advantage is that a specialist's high qualification allows for high machining accuracy, as the specialist can consider inaccuracies in the machine, tool, workpiece, and other factors. The method's disadvantages are a chance of rejection due to employee inattention, low productivity due to time spent on trial runs, and high cost due to the high qualifications of workers and the significant labor intensity.

The method of automatic accuracy on a pre-adjusted machine tool is based on a prior positioning of tools by an adjuster or a worker, ensuring that the specified size can be obtained in a single pass. After that, a batch of workpieces is machined by positioning the workpiece on the machine tool, switching it on, and making auxiliary movements, if necessary (bringing the tool to the workpiece, retracting it, etc.).

The method is highly accurate and productive and ensures dimensional stability for the batch of workpieces, but at high accuracy, it requires frequent adjustment and constant monitoring of machining parameters. In this case, the machining accuracy is influenced by subjective factors, such as the operator's qualifications, and objective factors, such as the state of a technological system and setup error. This method is used in almost all types of production. Machining workpieces on CNC machines, using the method of achieving accuracy with the help of limbs, using stops on universal machines, using semi-automatic and automatic machines, and copying machines—all these are examples of the method's implementation for achieving accuracy. The following methods can be used to adjust the position of the machine tools to the required size.

A method of using machine limbs suggests that when machining the first workpiece, a worker, using the previously discussed method of trial runs and measuring, determines limb positions of tool working bodies during machining the first workpiece, at which the accuracy of measurements is achieved. Subsequent workpieces are machined at the exact positions of the tool's limbs, ensuring the achievement of the operating dimensions. The limbs pre-setting to size can also be performed using standards or templates. The method is used in small-batch and medium-batch manufacturing. The accuracy depends on the following objective and subjective factors:

- objective—dependent on a machine (accuracy of a gauge, degree of machine wear, backlash in screw pairs, etc.);

- subjective—errors in the initial determination of limb readings (visual acuity of a worker, viewing angle relative to limb distributions, errors in pre-setting a limb to its original position, and other factors).

A method of tool setting during machining of a workpiece sample batch involves the following. Its core is based on achieving the required tool position by adjusting it based on the machining results of subsequent workpieces from a sample batch (5–10 pieces). This method of tool setting is similar to the method of trial runs and measuring. The method is used in medium-batch, large-batch, and mass manufacturing when adjusting turret machines and automatic and semi-automatic machines.

The essence of a reference method involves setting tools using a standard as a reference. The reference is a copy of a workpiece machined in a given operation. The reference dimensions include a probe thickness through which the tool touches the reference. The probe thickness also considers elastic deformations of machine elements occurring during operation due to cutting forces.

An adjustment method is based on monitoring parameters by sensors built into a technological system. Sensor signals are transmitted to devices that analyze them. These devices determine whether the actual dimensions match the specified dimensions and, if not, generate control signals to correct a tool position. This method is sometimes referred to as active accuracy control, which helps prevent defects, as opposed to passive control, which only registers them. In this case, the dimensional accuracy depends on the control device's and machine tool's accuracy.

This method is mainly used in large-batch and mass manufacturing, as it has high productivity and accuracy and reduces the time of control operations, but it requires additional costs for implementation.

4.3 Roughness

Geometric characteristics and physical and mechanical properties of a surface layer determine the surface quality of a machine part (Vukelic et al. 2022). A tangible surface borders the part and separates it from the environment (Petrakov et al. 2023). A nominal surface is an ideal surface with a nominal shape specified in a drawing or other technical documentation. Profile deviations of a tangible surface from the nominal one determine the geometrical characteristics of machined surface quality. These deviations include roughness, waviness, and deviations from the correct geometric shape. The latter was discussed earlier. Terms and definitions for surface roughness parameters are described in the standard (DSTU Gost 25142:2009 2009).

A surface roughness is a set of surface irregularities with relatively small steps, distinguished by a basic length (Fig. 4.10). A basic length l is the length of a basic line used to select irregularities characterizing the surface roughness.

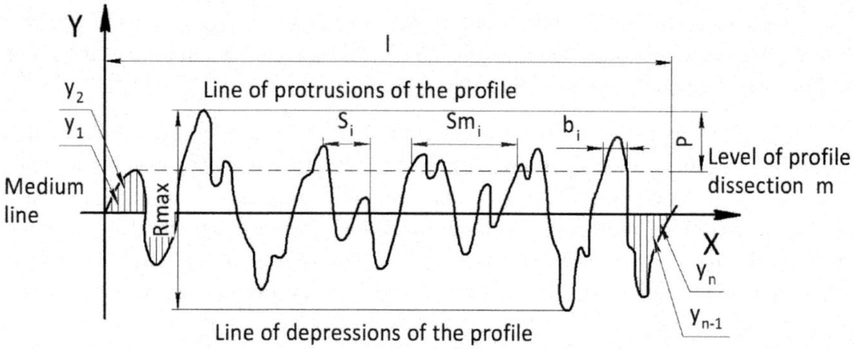

Fig. 4.10 Elements of surface roughness

The basic line (surface) is a line (surface) of a given geometric shape drawn in a certain way relative to a profile (surface), estimating geometric parameters of the surface. Surface roughness parameters are determined from a single datum, represented by a centreline m. A centreline *m* is a basic line with the shape of a nominal profile and is drawn to ensure the minimal standard deviation of the profile to this line within a center length.

A surface roughness is evaluated on a length *L*, which may contain one or more basic lengths (*l*). Values of the basic length are taken among the range: 0.01; 0.03; 0.08; 0.25; 0.80; 2.5; 8; 25 mm.

A profile deviation is a distance between any point on a profile and a centerline. Parameters and characteristics of surface roughness were specified in DSTU GOST 25142:2009 (2009).

A profile protrusion line is an equidistant line to a centerline and passes through the highest point of the profile within a basic length. A profile cavity line is an equidistant line to the centerline and passes through the lowest point of the profile within a basic length.

There are six parameters of surface roughness:

1. ***Ra***—an arithmetic mean profile deviation—absolute values of the arithmetic mean of profile deviations within a basic length:

$$Ra = \frac{1}{l}\int^{1} |y(x)|\mathrm{d}x; \quad Ra \approx \frac{1}{n}\sum_{i=1}^{n} |y_i|, \tag{4.1}$$

where

l basic length;
n the quantity of chosen points at the basic length.

2. ***Rz***—a height of profile irregularities by ten points—a sum of average absolute values of heights for the five most extensive profile protrusions and depth of the five most extensive cavities within a basic length:

$$Rz = \frac{\sum_{i=1}^{5} |y_{pi}| + \sum_{i=1}^{5} |y_{vi}|}{5}, \tag{4.2}$$

where

y_{pi} height of the i most extensive profile protrusion;
y_{vi} depth of the i most extensive profile cavity.

3. $\boldsymbol{R_{max}}$—the greatest height of profile irregularities—a distance between a profile protrusions line and a profile cavities line within a basic length l.
4. ***Sm***—an average spacing of profile irregularities—an average value of a profile irregularities spacing within a basic length:

$$Sm = \frac{1}{n}\sum_{i=1}^{n} S_{mi}, \tag{4.3}$$

where

n a number of spacings within a basic length l;
S_{mi} a spacing of profile irregularities equal to the length of a centerline segment limiting a profile irregularity.

5. ***S***—an average spacing of local profile protrusions—an average value of spacing of local profile protrusions within a basic length:

$$S = \frac{1}{n}\sum_{i=1}^{n} S_i, \tag{4.4}$$

where

n a number of spacings of irregularities along asperity tips within a basic length l;
S_i a spacing of profile irregularities along asperity tips equal to the length of a centerline segment between the projections of the two highest points of neighboring local profile protrusions onto it.

6. ***tp***—a relative reference length of a profile—a ratio of a profile reference length η_p to a basic length l:

$$t_p = \eta_p / l \tag{4.5}$$

A reference profile length η_p is a sum of segments length b_i within the basic length, cut off at a given level P in a profile material by a line equidistant to a centerline m:

$$\eta_p = \sum_{i=1}^{n} b_i. \tag{4.6}$$

Fig. 4.11 Designations of the surface roughness

It is recommended that the *Ra* parameter can be given preference in practice since samples for determining surface roughness by the comparison method are made using this parameter. The roughness parameter values and their designations and rules for applying them to product drawings are given in DSTU GOST 2512:2009 (2009).

To indicate the surface roughness with an unspecified type of machining, use the sign shown in Fig. 4.11a.

To indicate the surface roughness formed by removing a material layer (turning, milling, drilling, grinding, polishing, etching, etc.), use the symbol shown in Fig. 4.11b. This symbol should be used in such technological documentation as sketch sheets.

To indicate the surface roughness formed without removing a material layer (casting, forging, bulk stamping, rolling, drawing, etc.), use the symbol shown in Fig. 4.11c. The same sign indicates surfaces not subject to machining, i.e., remain in the state of the original workpiece.

4.4 Surface Quality Impact on Exploitation Properties

The relationship between surface quality parameters of parts and their operational properties is one of the main areas of research in machine building and instrumentation.

Currently, the relationship between the machined surface quality and important performance characteristics of machine parts and components has been sufficiently studied (sliding and rolling friction and wear, liquid friction, contact hardness, strength of press connections, reflectivity, wear resistance under variable loads, corrosion resistance and quality of paint and varnish coatings, measurement accuracy, the correlation between dimensional tolerances and surface roughness, etc.). Some physical and chemical properties of a surface layer do not affect the performance characteristics of machine parts; instead, their chemical composition and structure (microstructure) are crucial (Berladir et al. 2022).

Part friction and wear are significantly related to macro- and microirregularities, waviness, and the direction of the machining strokes (marks).

The characteristics of macro-irregularities and waviness affect the size of the areas with actual contact, i.e., determine the contour of the contact area. The presence of waves reduces the contact area by 5–10 times compared to a flat, rough surface. The wave height *Wz* is more critical than the wave spacing *S* because the former parameter has a more substantial effect on the size of the contact area and, as a result, on wear (Fig. 4.12). The macro-irregularity's shape and size largely determine the process of surface contact. The limit values of deviations from the correct geometric

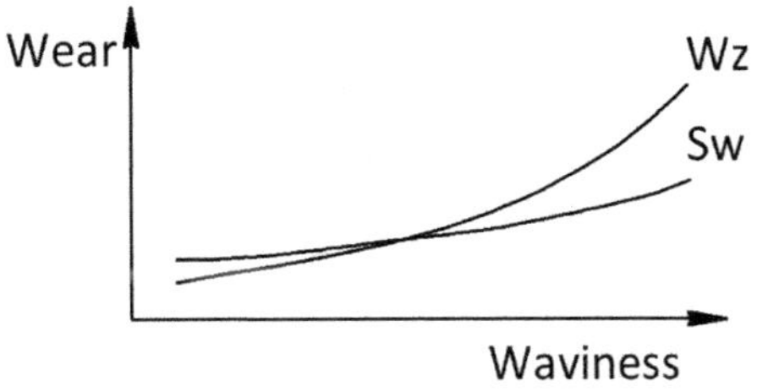

Fig. 4.12 Height W_z and wave spacing S_ω effect on wear

shape and the relative location and shape of macro-irregularities of the surfaces of the parts in contact must be considered when assessing the macro-irregularities effect.

The mutual movement of contacting planes (Fig. 4.13a) or cylindrical (Fig. 4.13b) surfaces with macro-irregularities (roughness) causes shearing, fragmentation, and plastic displacement of the irregularity vertices at the start of operation because of their contact along the irregularity vertices.

The correlation between wear and the operating time of frictional areas is shown in the diagrams (Fig. 4.13d, e). Initial wear occurs relatively quickly (area I) within period T1. The correct lubrication regime (Fig. 4.13c) results in slower wear (area II) due to the generation of equilibrium roughness. This period determines the service life of the part. Area III demonstrates the catastrophic wear of the pair.

Curve *2* (Fig. 4.13e) represents the wear of surfaces with lower initial roughness than Curve *1*. In this case, the amount and run-in time are reduced while the intensity of operational wear remains the same. The service life of the friction pairs within the size A of the permissible wear will vary. With a lower roughness of the connected surfaces, the service life of the parts will be longer ($T_2 > T_1$).

The initial wear (run-in) process relief is transformed into an operational relief (Fig. 4.14a)—consequently, the size and shape of irregularities and the direction of machining marks change. The actual contact area of the surfaces increases as the profile's reference length *tp* increases (the reference surface curve is given in Fig. 4.14b).

The height of the irregularities decreases or increases to some optimum value, which varies for different conditions. It has been experimentally established that the lowest wear occurs not at the minimum roughness of the frictional areas but at the roughness with the optimum value of R_{opt}, with deviations to the left or right leading to an increase in wear (Fig. 4.15, Curve *1*). Under heavier operating conditions, wear Curve *2* shifts to the right and upward, and the points of optimum roughness shift to the right toward increasing the height of the irregularities.

A height increase in irregularities compared to the optimum value increases wear due to higher mechanical adhesion, chipping, and shearing of irregularities. A height decrease of irregularities compared to the optimum value drastically increases wear due to molecular adhesion and surface seizure, facilitated by lubricant squeezing and, thus, improper lubrication of mirror-clear surfaces. That is why scraped surfaces are preferable to lapped ones, as they have grooves ("pockets") that retain the lubricant. A layer of porous chromium ensures good lubricant retention, the porous structure

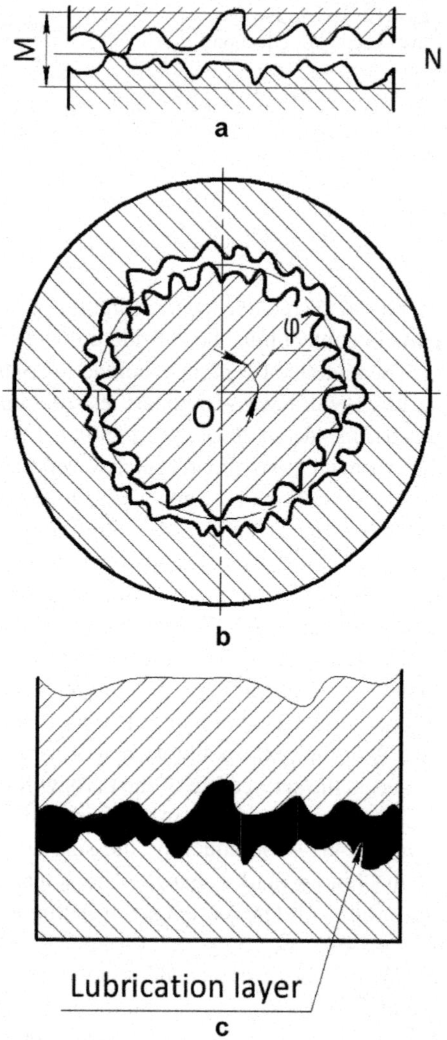

Fig. 4.13 Effect of surface roughness on machine parts wear: **a**, **b**—contact schemes of joined parts along the generatrix (along the axis) and circularly; **c**—virtual and physical contact of surfaces; **d**, **e**—typical wear schedules over time

of metal-ceramic parts, and a system of small lubricant retention channels obtained by vibration rolling.

An optimum roughness is characterized by the height, spacing, and shape of irregularities (cavity radius, the inclination angle of irregularities in the direction of movement, etc.). The optimum roughness parameters depend on the frictional areas' lubricant quality, other operating conditions, and their design and material. The range of R_{opt} is usually minimal. Pointed microirregularities wear out faster than flat microirregularities (Fig. 4.16) since the smaller contact area.

Fig. 4.13 (continued)

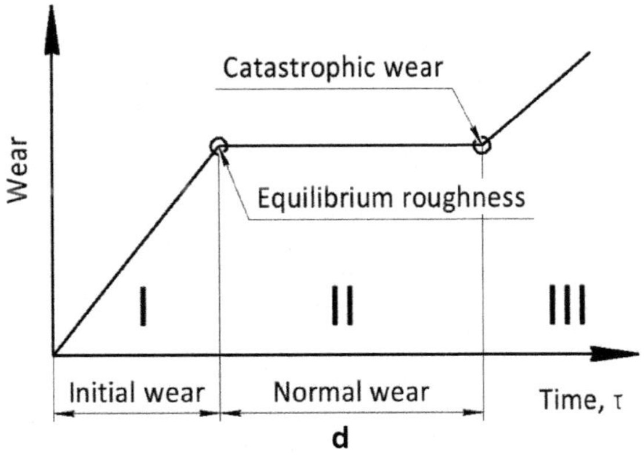

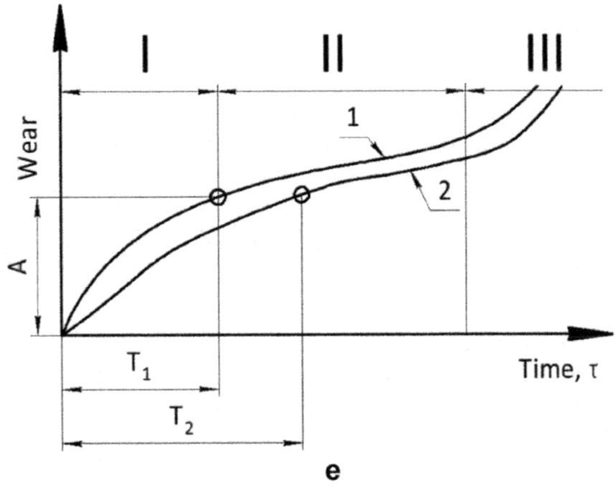

The microhardness of a surface layer affects wear resistance. Preliminary deformation hardening (riveting) of metal in this layer reduces surface crushing and abrasion when surfaces are in contact. For example, deformation hardening resulting from cutting reduces surface wear by 1.5–2 times. The positive effect of preliminary deformation hardening on parts' wear resistance is noticeable not only under friction with a lubricant but also to the same extent under dry friction: wear resistance increases by 1.5–2 times or more. The effect of deformation hardening on wear resistance is powerful for more ductile and relatively mild steels, for which even a slight increase in microhardness causes a significant reduction in wear.

Frictional areas acquire optimum roughness and form optimum microhardness of metal in the upper layer during the run-in. Deformation hardening positively affects the wear resistance of frictional areas only up to a specific value. At high microhardness, wear increases due to "overriveting" due to the flaking of metal particles. Therefore, the metal hardening of a surface layer during the machining of

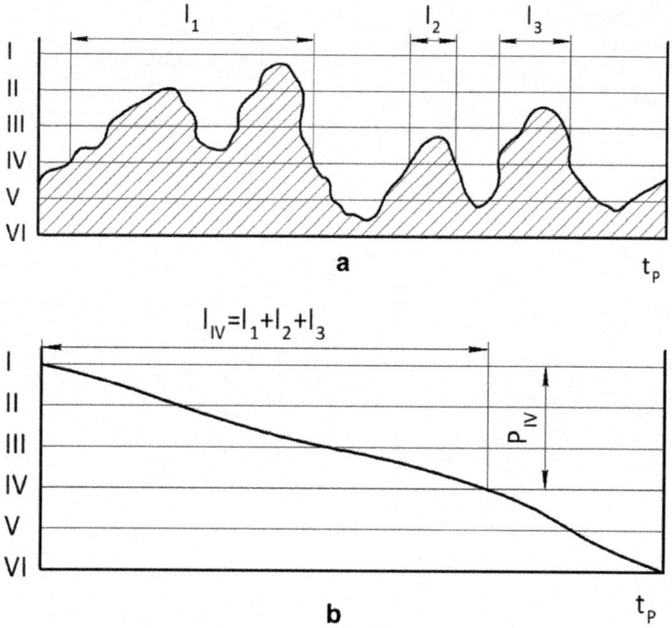

Fig. 4.14 Process relief transformation into an operational relief

Fig. 4.15 Wear correlation with surface irregularity vertices and operational conditions

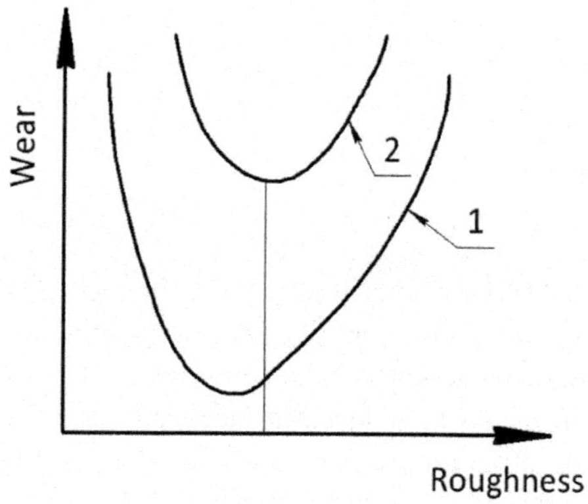

parts using special hardening operations should be performed at a strictly regulated value of deformation hardening to prevent the occurrence of "overriveting".

Wear is significantly reduced by thermal and chemical-thermal treatment of parts (e.g., surface hardening, cementation, cyanidation, nitriding, chromizing, boronizing, aluminizing, siliconizing, etc.), padding and plasma spray of parts with hard alloys, as well as by galvanic application of hard coatings (chrome-plating).

A surface layer's hardness, structure, and chemical composition affect the wear reduction. Residual compressive stresses in the layer reduce wear slightly, while

Fig. 4.16 Contact area depending on an irregularity shape: **a**—pointed; **b**—flat-topped; r, r′—radii of irregularities vertices, S—irregularity spacing

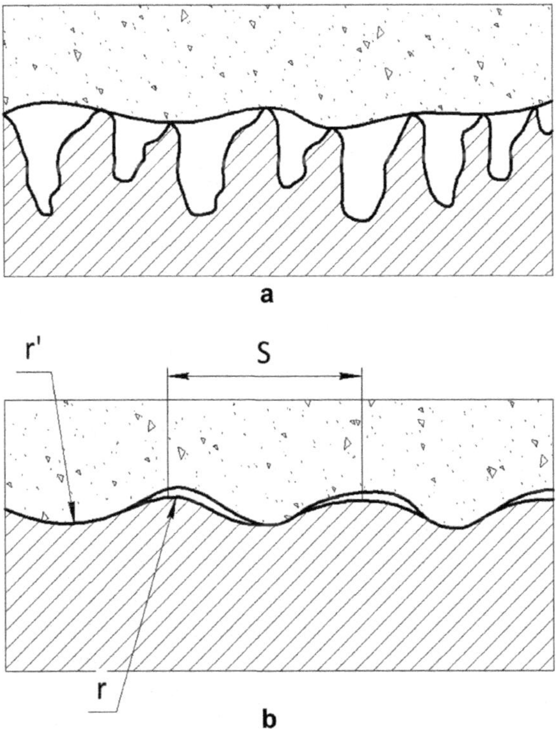

residual tensile stresses increase it. This effect is more evident in elastic contact and less evident in elastic–plastic contact. Wear changes the residual stresses in the surface layer of the part. Parts' strength also depends on the surface roughness. Failure of a part, especially under variable loads, is mainly caused by the concentration of stresses due to irregularities. The lower the roughness, the lower the risk of surface cracks due to metal fatigue. Finishing parts (e.g., finishing, polishing, etc.) significantly increases fatigue strength.

Reducing surface roughness significantly increases the corrosion resistance of parts. This is especially important when protective coatings are not used for surfaces (engine cylinder surfaces, etc.).

A directional pattern of strokes—traces of machining and other machining—is an important geometric characteristic of surface quality. It affects the surface's wear resistance, fit accuracy, and press joint strength. In critical cases, the designer should specify the directional pattern of machining marks on the part's surface. For example, this factor may be necessary in relation to the direction of relative sliding of connected parts or the movement of a liquid or gas jet through the part.

Wear is reduced and minimized when the sliding direction coincides with the direction of the irregularities in both parts.

Surface roughness and waviness are interdependent with dimensional accuracy because the connection accuracy, determined by the gap size in a joint, depends mainly on the ratio of irregularities height and tolerance field (machining accuracy) of each connected part.

Given that during the initial wear, the height of irregularities can decrease by 65–75% (at a higher height than at the optimum roughness), an increased gap appears in the joint, which can reach the tolerance value for the part manufacturing, and the accuracy of the joint will be compromised entirely (for example, instead of the 6th accuracy class joint required by the drawing, a 7th or 8th accuracy class joint is formed; instead of a tension fit, transitional fit appears, etc.). Parts' machining until the optimum friction surface roughness is achieved helps prevent this in all critical joints, requiring long-term preservation of the accuracy specified by a design engineer (Jovicic et al. 2023).

Surface roughness and waviness are always essential for high accuracy, determined by the operating conditions of connected parts and the need for reliable measurement results in production (Halchuk et al. 2023). The reduction in surface roughness adds additional accuracy to the nature of the joint because the gap (or tension) obtained from the inspection of parts differs from the effective gap or tension that occurs during assembly and then in operation. The effective tension during assembly decreases, and the gap increases more rapidly during the mechanism's operation, in contrast to more roughly machined connected surfaces.

The low surface roughness can also be used to enhance the appearance of a workpiece or to facilitate the cleaning of surfaces during operation.

References

Berladir K, Trojanowska J, Ivanov V, Pavlenko I (2022) Materials selection in product development: challenges and quality management tools. In: Hamrol A, Grabowska M, Maletič D (eds) Advances in manufacturing III. MANUFACTURING 2022. Lecture notes in mechanical engineering. Springer, Cham. https://doi.org/10.1007/978-3-031-00218-2_7

DSTU 2498–94 (1994) Basic interchangeability standards. Tolerances of form and position of surfaces. Terms and definitions. Derzhstandart of Ukraine, Kyiv (in Ukrainian)

DSTU GOST 25142 (2009) Surface roughness. Terms and Definitions; derzhspozhivstandart of Ukraine, Kiev 2009 (In Ukrainian)

Halchuk TN, Povstyanoy OY, Bembenek M, Redko RG, Chetverzhuk TI, Polinkevych RM (2023) Impact of technological system's characteristics on the machining accuracy of bearing rings. J Eng Sci 10(1):A22–A30. https://doi.org/10.21272/jes.2023.10(1).a4

Jovicic G, Milosevic A, Kanovic Z, Sokac M, Simunovic G, Savkovic B, Vukelic D (2023) Optimization of dry turning of inconel 601 alloy based on surface roughness, tool wear, and material removal rate. Metals 13:1068. https://doi.org/10.3390/met13061068

Karpus VE, Ivanov VA (2012) Locating accuracy of shafts in V-blocks. Russ Eng Res 32:144–150. https://doi.org/10.3103/S1068798X1202013X

Kushnirov P, Denysenko Y, Pop G, Basov B, Dynnyk O (2024) Stiffness of underpinning supports for fixtures. In: Ivanov V, Trojanowska J, Pavlenko I, Rauch E, Piteľ J (eds) Advances in design, simulation and manufacturing VII. DSMIE 2024. Lecture notes in mechanical engineering, Springer, Cham, pp 251–262. https://doi.org/10.1007/978-3-031-61797-3_21

Petrakov Y, Danylchenko Y, Sapon S, Sikailo M (2023) Surface Relief formation in peripheral end milling. In: Ivanov V, Trojanowska J, Pavlenko I, Rauch E, Piteľ J (eds) Advances in design, simulation and manufacturing VI. DSMIE 2023. Lecture notes in mechanical engineering, Springer, Cham. pp 316–326. https://doi.org/10.1007/978-3-031-32767-4_30

Petrakov YV, Ohrimenko OA, Sapon SP, Sikailo MO, Fedorynenko DY (2023) Determination of chatter-free cutting mode in end milling. J Eng Sci (Ukraine) 11(2):A1–A11. https://doi.org/10.21272/jes.2024.11(2).a1

Pop GM, Crisan LA, Tripa M (2020) The effectiveness of ICT tools for engineering education: ISO checker. In: Ivanov V et al (eds) Advances in design, simulation and manufacturing II. DSMIE 2019. Lecture notes in mechanical engineering, Springer, Cham, pp 270–279. https://doi.org/10.1007/978-3-030-22365-6_27

Pop GM, Comes R, Crisan LA (2021) An android application for explaining form deviations using 3D models. In: Ivanov V, Trojanowska J, Pavlenko I, Zajac J, Peraković D (eds) Advances in design, simulation and manufacturing IV. DSMIE 2021. Lecture notes in mechanical engineering, Springer, Cham, pp 311–321. https://doi.org/10.1007/978-3-030-77719-7_31

Pop GM, Comes R, Neamtu C, Crisan LA, Andresan D (2023) Geometrical tolerances android application developed for student for a better understanding of orientation tolerances according to the latest ISO standards. In: Durakbasa NM, Gençyılmaz MG (eds) Towards industry 5.0. ISPR 2022. Lecture notes in mechanical engineering, Springer, Cham, pp 3–15. https://doi.org/10.1007/978-3-031-24457-5_1

Vukelic D, Simunovic K, Kanovic Z, Saric T, Doroslovacki K, Prica M, Simunovic G (2022) Modelling surface roughness in finish turning as a function of cutting tool geometry using the response surface method, Gaussian process regression and decision tree regression. Adv Product Eng Manag 17(3):367–380. https://doi.org/10.14743/apem2022.3.442

Yakovenko IE, Permyakov OA, Fesenko AV (2022) Technological basics of machine building. NTU «KhPI», Kharkiv

Chapter 5
Technological Processes in Machining

5.1 Fundamental Principles

To design technological processes, such fundamental principles should be observed: technical (quality of products); economic (minimal costs to make products); and ecological (minimal damage to workers and the environment).

All initial information for designing technological processes can be subdivided into three groups: basic (drawings of machines, assembly units, parts; output including spare parts; production time; manufacturing system data—ways of making workpieces, etc.); guideline (work safety instructions, conventional technological processes, standards of designing and adopting technological processes, etc.); and reference (catalogs of equipment, fixtures, machining modes, workpiece time, etc.).

When designing technological processes, basic and reference data must reflect the actual state of the manufacturing system. Catalogs of new equipment and fixtures may be used for promising technological processes.

The design of machining technological processes is a multi-variant task (Kusyi and Stupnytskyy 2020). It implies processing initial information (insufficient and inaccurate) into end-of-end technological solutions. The design itself has many stages. It synthesizes new data at each stage, which supplements the initial data. Calculations should be optimized to make a multi-variant solution (e.g., to define ways of surface machining or equipment selection) (Petrakov et al. 2023). The design of the technological process is iterative. In other words, you can return to and correct previously made solutions.

Within the design of the technological process, there are three large stages (Fig. 5.1): collecting and analyzing primary information (green color); making technological solutions (blue color); and providing a technical–economic evaluation and adopting the technological documentation (red color). The design sequence of these stages may be represented as the below-mentioned algorithm (Zakharkin 2004).

There are such design methods of technological processes: manual, computer-aided, and automatic.

V. Ivanov et al., *Fundamentals of Manufacturing Engineering Using Digital Visualization*, Springer Tracts in Mechanical Engineering,
https://doi.org/10.1007/978-3-031-74360-3_5

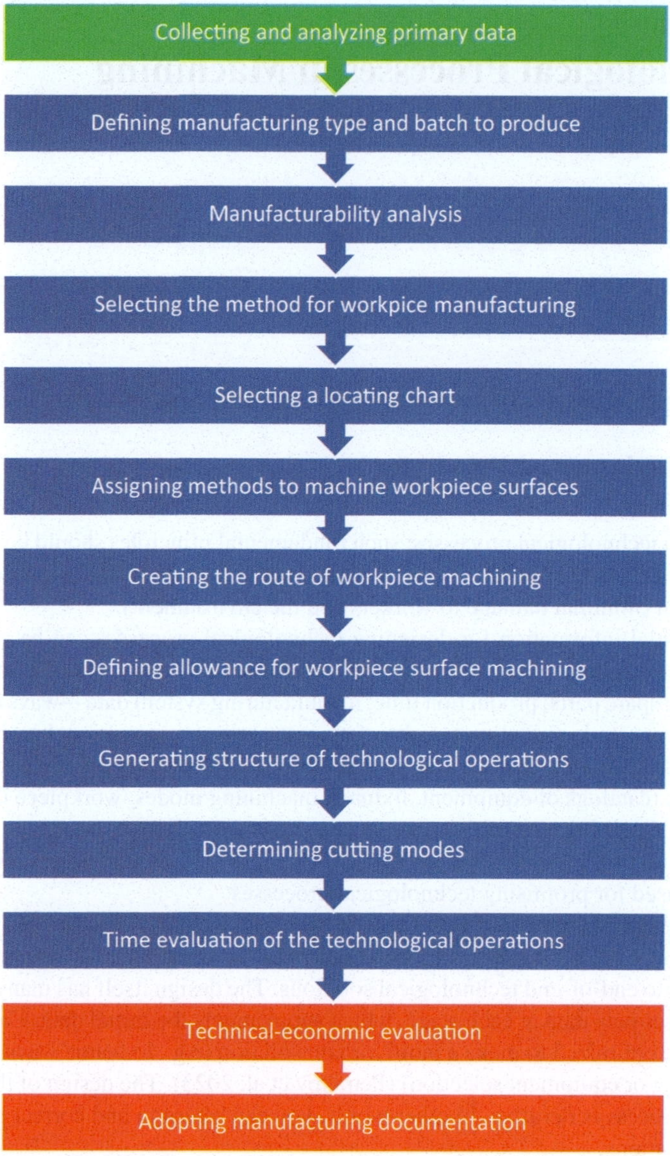

Fig. 5.1 Algorithm of technological process design

The manual design means that all object descriptions (drawings, manufacturing documents, etc., in different languages) are conducted by humans. Via intuition and experience, they settle most stages: analyzing technical conditions, selecting ways of workpiece locating and production, and assessing the sequence of operations. Here, you observe standards and recommendations, follow typical solutions, and make some technological calculations. The computer-aided design is implemented within the interaction between computers and humans using the specialized computer-aided process planning (CAPP) software. The automatic design is carried out without humans at all.

In the case of CAPP, computers mainly solve tasks whose essence may be formalized and arranged as functional and structural data: calculations, stages of machining part surfaces, optimal conditions to decrease laps, etc. Further development of design and computing theory may provide computers with creative tasks to settle.

Modern CAD systems solve technological tasks via quick communication between humans and computers. It is efficient for creative tasks with the heuristic approach (e.g., recognition of part images, their size and topology to make an optimal choice of locating, machining, assembling schemes, etc.). These and other tasks can be settled effectively by synthesizing creative human skills and "software abilities".

In terms of design principles, there are two methodological approaches. The first is synthesizing the technological process via knowledge and solutions derived from primary definitions (step, setup, compatibility of machines and tools, etc.). The second includes typical and group technological processes and integrated solutions accumulated because of a long engineering experience.

Technological process synthesis is a difficult, laborious, and formal process. It requires deep knowledge, intuition, and experience (Kusyi et al. 2022).

The typical or group technological process design does not need deep knowledge. It is easily formalized and automated. Today, almost all valid technological process design systems (computer-aided and automatic) are based on typical technological solutions.

Defining the type of manufacturing, output tact, and batch is essential at the technological process's design stage. There are three types of manufacturing: single-part, batch, and mass. For a specific product, the type of manufacturing can be established through tables in specialized literature. Here, the type of manufacturing is a function between product output and their mass/total labor intensity/dimensions. For the large-batch and mass manufacturing, output tact must be calculated. The optimal number of products launched into production simultaneously should be established for small-batch and medium-batch manufacturing. It is called an $N_{part\ batch}$ or $N_{machine\ batch}$. The batch output of parts is estimated via the formula:

$$N_{partbatch} = \frac{N \cdot a}{254} \text{ pieces,} \tag{5.1}$$

where

N	annual output of parts;
$a = 1, 2, 5, 10, 20$	number of days when a stock of parts is required;
254	number of workdays per year.

If the annual output is evenly shared per month, $N_{machine\ batch}$ is usually equal to the monthly output plan:

$$N_{machinebatch} = \frac{N}{12}\text{ pieces} \quad (5.2)$$

There are other ways to calculate these indexes as well. Sometimes, the annual output share among months may be defined by a customer.

One of the factors that significantly influence technological processes is testing product construction for manufacturability. The product construction must not only correspond to its service requirements but also to a more economical manufacturing. The lower labor intensity and product cost are, the more technological it is. Therefore, the manufacturability test is performed before the technological process design. It demands deep manufacturing knowledge, including technical and economic abilities, equipment, etc. Technological product construction should include a wide use of unified assembly units, standard parts, and their elements, as well as a minimal quantity of unique parts. The technological product construction should correspond to assembly requirements. Besides, it is necessary to have convenient assembly bases, minimal adjustment, and opportunities for parallel assembly of separate units.

Testing product construction for manufacturability can be qualitative and quantitative. The qualitative testing characterizes manufacturability in a generalized form. Based on the tester's experience, it is conducted on all design stages as a preliminary condition. It is assessed via values "good—bad". Quantitative testing defines the relation of achieved indexes to basic ones. As basic, indexes of those machines are regarded as belonging to the best products worldwide.

Manufacturability must be tested in the following way: selecting and analyzing initial materials needed for manufacturability assessment; defining manufacturability indexes of the basic and producible construction; conducting the comparative assessment and calculation of manufacturability; and developing measures to raise manufacturability.

The main indexes of product construction manufacturability are labor intensity, cost, material, and energy consumption.

The labor intensity of product assembly, manufacture, or repair is the sum of standard hours used for its assembly, manufacture, and repair. The level of construction manufacturability as to labor intensity is defined via the formula:

$$C_m = \frac{L_e}{L_b}, \quad (5.3)$$

where

L_e and L_b expected (design) and basic labor intensities to manufacture or repair products (standard hours).

Cost is the total expenditure value for materials and salaries with extra fees and overhead expenses. That comprises all components of the product. Cost is a generalized index of product quality. The level of construction manufacturability as to cost is defined via the formula:

$$C_c = \frac{C_e}{C_b}, \tag{5.4}$$

where

C_e and C_b expected (design) and basic costs to manufacture products (euro).

Material consumption implies all materials used to manufacture a product of a certain mass. This is often regarded as a ratio of product mass to one of its technical parameters (e.g., capacity).

Energy consumption refers to the quantity of fuel and power resources applied in product manufacturing. It can be measured in kilowatts, calories, etc.

Levels of construction manufacturability in terms of material and energy consumption are defined similarly to those in terms of labor intensity and cost.

The construction is technological if the values of manufacturability are lower than one. There are also other indexes to clarify construction drawbacks and raise manufacturability. They comprise the unification of parts and their components, material brands, assortments, thread or fit size, etc. Figure 5.2 shows non-technological solutions on the part construction: the opening axis is not perpendicular to the surface, which makes it necessary to design and produce a special drill. Another example (Fig. 5.2b) represents a more accurate opening (the seventh finish degree) as blind. Here, we cannot cut such an opening. On the contrary, other progressive machining methods can be used. In such a case, it is reasonable to make a thorough opening.

After the manufacturability test, all recommendations about construction change should be described in the explanatory report. The change proposals should be included in the product construction if they do not contradict the product service.

Selecting the method for workpiece manufacturing is an important task. Primary workpieces are widely applied in engineering. They are obtained via casting, plastic straining, structural sections, welding, powder metallurgy, etc. Such ways of workpiece manufacturing may be implemented via various methods. For example, casting is promoted via molds. Plastic straining may be realized in the cold and hot states with or without clichés (free forging). Workpieces may have different configurations of structural sections (round, sheet, or complex).

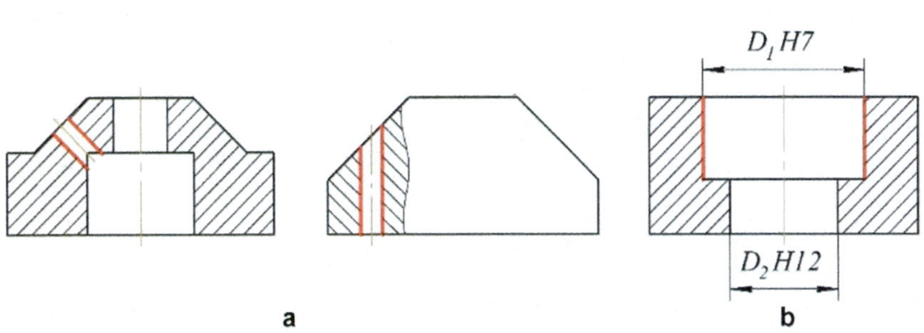

Fig. 5.2 Examples of non-technological part solutions

Manufacturing of primary workpieces must be carried out, including such features: type of manufacturing, part material, part dimensions, complexity, wall thickness, stepped shafts, and opening diameter in case of casting or plastic straining.

The more accurately the workpiece is made, the more its form and size are closer to the finished part parameters. Simultaneously, this reduces the time needed for machining and raises costs.

Besides, the enterprise potential of workpiece manufacturability should be considered. Otherwise, workpieces must be produced at other cooperating enterprises, which increases expenditure.

In each case of selecting ways to make a primary workpiece, you should conduct an economic analysis to determine the most reasonable approach. Here, the following ratio must be observed:

$$C_{total} = C_{wm} + C_l \rightarrow min \tag{5.5}$$

where

C_{total} total cost of the workpiece and its machining;
C_{wm} the cost of workpiece manufacturing itself;
C_l the cost of extra machining to remove laps, etc. (comparison of several alternative variants).

The machining essence consists of removing superficial layers of workpiece materials to ensure the workpiece surface's proper size, form, and quality. The removed layer determines part quality, labor intensity, coefficient of workpiece material use, technological process reliability, cost, etc. Allowance (Z) is the material layer that should be removed from the workpiece surface in case of its machining. The unmachined surfaces do not have any allowance. The workpiece surface allowance may be removed during one or several operations. Intermediate allowance (Z_i) is the material layer that must be removed during one operation (step). Total allowance (Z_{total}) is the material layer removed from the workpiece during all operations to acquire a proper form, size, and quality of the superficial part layer.

Depending on the workpiece location, there are asymmetrical and symmetrical allowances. The former are removed consecutively from one or opposite surfaces. The latter are situated on inner and outer rotation surfaces and on both opposite surfaces, machined simultaneously.

Sometimes, it is unreasonable to manufacture complex workpieces because of economic and technical reasons. The workpiece form significantly differs from the finished part form in such cases. The material added for form simplification during manufacturing is called lap.

The primary data used to calculate laps can be part drafts or drawings with all dimensions and requirements for accuracy and quality of the superficial layer, the kind and way of primary workpiece manufacture, and the route of part manufacture (surface machining).

There are such methods of defining laps: statistical research; analytical calculation; technological dimension chain.

The statistical research implies that total and operational allowances are taken from tables (Lieposhkina et al. 2023). They are based on the generalization and systematization of manufacturing data from advanced enterprises. Such tables belong to state workpiece standards. The method drawback is that allowances do not include concrete conditions of technological process planning: no machining route (total allowance), no workpiece location scheme, and no previous machining errors (intermediate allowance). Within statistical research, allowance dimensions are overrated. They are directed at machining conditions when you should avoid any defective products with the allowance of use.

5.2 Strategy for Workpiece Machining Routes

The workpiece machining route defines a sequence of cutting, thermal, galvanic, fitting, and controlling operations (Kusyi et al. 2024). Typical, group or working factory technological processes can be used for route design.

In the case of the technological process route design, it is necessary to consider the factory experience and recommendations of literature sources to divide the technological process into stages properly. They combine methods with approximately similar accuracy and quality of surface machining.

One of the schemes to represent a reasonable sequence of workpiece machining stages is shown in Table 5.1. It generalizes a long engineering experience. According to the table, finished technological datums are first, and other surfaces are second to a machine. This is conducted to adjust the surfaces to meet the part drawing requirements. The highest accuracy belongs to part surfaces that perform service functions. Therefore, the workpiece machining route must be subordinated to one of the main principles: product service use. Such a route significantly influences the technological process sequence.

As the table indicates, the stages are optional for all technological processes (only some parts demand heating, surfacing, and finishing). For precise workpieces, it is unnecessary to include rough, semi-final, and final machining stages. The optional stages can be skipped when generating a machining route.

Outer surfaces have maximal accuracy and minimal undulation. Consequently, the manufacturing route must be ended with part finishing. Other surface treatment is completed in the semi-final or preliminary machining stages.

Division of the technological process into stages provides many advantages. The preliminary treatment may be done using old or inaccurate equipment by less skilled specialists. The time gap between preliminary, semi-final, and final machining reveals more deformations to remove during the last treatment stage. Finishing at the route end decreases the surface damage risk. There may be deviations from the main principles of the workpiece machining route generation. Thus, the route often ends in treating damageable surfaces (outer threads, etc.). For unsmooth parts, it is sensible to conduct final machining immediately after rough and preliminary machining to find internal defects. This rule is usually observed when machining flat surfaces

Table 5.1 Sequence of workpiece machining stages

Stage	Name	Features
1	Preliminary I	Treatment of surfaces that will be applied as technological datums in further stages
2	Preliminary II	Rough machining of outer surfaces and those that do not presuppose defects. Size accuracy: IT12–IT14. Forms and location accuracy: degrees X–XII. Roughness $R_z = 10–20\ \mu m$. $R_a = 2.5–5\ \mu m$
3	Thermal I	Thermal treatment to remove inner tension of kind I and II
4	Light	Correction of technological datums and semi-final surface machining. Size accuracy: IT10–IT12. Forms and location accuracy: degrees VIII–IX. Roughness $R_z = 6.3–10\ \mu m$. $R_a = 1.25–2.5\ \mu m$
5	Thermal II	Thermal treatment to improve the quality of outer material layers
6	Final	Correction of technological datums and final surface machining. Size accuracy: IT8–IT9. Forms and location accuracy: degrees VI–VII. Roughness $R_z = 3.2–6.3\ \mu m$, $R_a = 0.63–1.25\ \mu m$
7	Extra	Secondary operations (fastener drilling, beveling, grooving). Treatment of damageable surfaces (threading)
8	Galvanic	Chromizing, nickeling, etc
9	Expectant	Finishing of outer and zero surfaces. Size accuracy: IT5–IT7. Forms and location accuracy: degrees IV–V. Roughness $R_z = 0.8–1.6\ \mu m$, $R_a = 0.16–0.32\ \mu m$
10	Control	Final testing

via turning and drum mills. Here, you reach a higher treatment concentration and decrease part setup and machining costs. If parts are subjected to thermal treatment, the machining route is subdivided into several elements (Table 5.1).

It is worth saying that Stage 3 may be implemented as first and second, which allows not to break the machining sequence. Stage 5 corresponds to hardening and cementing.

In the case of nitriding, it can be done before Stage 8. Consequently, Stages 5 and 6 may be skipped.

Thermal treatment leads to errors in the workpiece form, mutual surface location, and undulation. To remove these defects, routes should correct technological datums or machine some surfaces again (Stages 4 and 6). Thermal treatment often includes specific operations, such as coppering uncementable surfaces. Technical control should be planned within the technological process design (Denysenko et al. 2022, 2023). It is realized to prevent defects (Silva et al. 2023). Besides, technical control is reasonable before important operations and at the end of treatment. For all other operations, random control is usually applied.

5.3 Selection of Machines and Fixtures

Workshop equipment should be analyzed to select a cutting machine (Antosz et al. 2022). Within single-part manufacturing, very loaded equipment that limits output may be used only in exceptional cases (when no other machining methods are applicable). Otherwise, any available equipment should be used that promotes high-quality part machining (Trojanowska et al. 2022). For technological process design at new enterprises, equipment selection is restricted only by economic considerations (Dobrotvorskiy et al. 2020).

The selection of machines is based on such criteria: type of manufacture; technological possibilities of realizing processing methods (structure of technological steps); working area dimensions; quantity of tools; engine capacity; machine price.

General-purpose and versatile machines are involved in small-batch and single-part manufacturing.

High-performance machines possess a restricted technological potential. However, their enhanced capacity and undulation promote better cutting. Such equipment comprises multi-tool lathes; hydro-copying, single-spindle and multi-spindle machines; odd or centerless grinding machines; turning and drum mills, etc. They are appropriate for the large-batch and mass manufacturing (Redko et al. 2019).

Special machine tools are based on high performance. They are equipped with extra spindles and other units (Krol et al. 2024) for specific machining operations in mass manufacturing.

Special machine tools are designed and manufactured for exclusive orders. They are used to perform certain operations. Both design and manufacture of such equipment are costly. Therefore, they are repaid only in mass manufacturing. Modular machines (Zaleta et al. 2023) are involved in batch and mass manufacturing. CNC machines are used for small-batch and medium-batch manufacturing. Equipment selection must be based on the technological process expenditure analysis within the product life cycle with a certain quality (Yakovenko et al. 2021). The analysis results are assessed via ratios between principal and artificial time and overall machining expenditure.

The most multi-variant machine is a lathe. Selection of lathes, single-spindle, and multi-spindle machines may be conducted without complicated calculations via diagrams. They define the economic use limit of these machines for different output programs. In each selection case, machine certificates are assessed. If they are absent, catalogs of cutting and other equipment are involved.

The following considerations define the selection of machining fixtures (Karpus and Ivanov 2012a). Universal fixtures are widely used in single-part and small-batch manufacturing (e.g., chucks, vises, dividing devices, rotary tables, etc.). Dedicated fixtures are applied for large-batch and mass manufacturing to reduce production time with a higher accuracy (Karpus and Ivanov 2012b). Fixture selection must be based on the technological process implementation expenditure analysis, including limited time and products. From this perspective, you should compare fixtures with similar requirements and tasks in certain conditions. The optimal choice of fixture must consider (Karpus' and Ivanov 2008): sufficient flexibility for the machining of parts within the limitations of the machine tool; specified accuracy of machining; mechanized or automated adjustment on switching to a new product; the rigidity of the components and subassemblies of the fixture to withstand considerable cutting forces and permit maximum utilization of the equipment's capacity; adequate tool accessibility to machine the maximum number of surfaces in a single setup; standardization of the parts and subassemblies of the fixture for reasons of affordability; high reliability of the fixture and its components; and economic efficiency (Ivanov et al. 2021).

In manufacturing engineering, there are such fixtures (Karpus et al. 2012). Universal assembled fixtures consist of all-purpose interchangeable elements. They are employed as special reversible fixtures of short-term functions. They locate and clamp workpieces within the dimensions of the fixture set.

Assembled-and-disassembled fixtures are made of standard elements with their additional mechanical processing. They are applied as special non-reversible fixtures of long-term action from reversible elements.

Dedicated fixtures are manufactured from universal standard parts and units. They are applied as non-reversible fixtures of long-term functions. Their elements are produced from non-reversible parts.

Universal non-adjusting fixtures are the most widely applied systems within batch manufacturing. They comprise lathe chucks, vises, etc. These devices fix small and medium workpieces (depending on their space location). At the same time, the setup of workpieces is connected with the need to control their orientation in space. Such fixtures ensure the performance of a wide range of workpiece machining operations.

Universal adjusting fixtures equipped with specialized adjustments ensure the setup and fixturing of small- and medium-size workpieces and the performance of a wide range of machining operations. They are used as fixtures of short-term functions.

Specialized adjusting fixtures provide locating and clamping of workpieces according to a specific locating chart using special adjustments to perform typical machining operations.

All of the fixtures mentioned above are unified. Their selection dramatically differs from that of conventional tooling. Fixture designers usually perform this work. Manufacturing engineers only define a proper fixture system and propose a locating chart.

Within fixture selection, the following documents should be applied: normative (standards of fixtures and their components, manufacture terms and definitions); technical (albums of typical fixture configurations, catalogs, specifications of equipment, and instructions on fixture selection).

References

Antosz K, Jasiulewicz-Kaczmarek M, Waszkowski R, Machado J (2022) Application of Lean Six Sigma for sustainable maintenance: case study. IFAC-PapersOnLine 55(19):181–186. https://doi.org/10.1016/j.ifacol.2022.09.204

Denysenko Y, Trojanowska J, Tsarytsyn V, Husár J (2023) Quality management of the process of the complex parts control. In: Knapčíková L, Peraković D (eds) 7th EAI international conference on management of manufacturing systems. MMS 2022. EAI/Springer innovations in communication and computing, Springer, Cham, pp 159–170. https://doi.org/10.1007/978-3-031-22719-6_12

Denysenko Y, Górski F, Ciszak O, Berladir K, Kushnirov P (2023) Quality management at the manufacturing enterprise: repair processes case study. In: Ivanov V, Trojanowska J, Pavlenko I, Rauch E, Piteľ J (eds) Advances in design, simulation and manufacturing VI. DSMIE 2023. Lecture notes in mechanical engineering, Springer, Cham, pp 93–101. https://doi.org/10.1007/978-3-031-32767-4_9

Dobrotvorskiy S, Basova Y, Dobrovolska L, Sokol Y, Kazantsev N (2020) Big Challenges of Small Manufacturing Enterprises in Industry 4.0. In: Ivanov V, Trojanowska J, Pavlenko I, Zajac J, Peraković D (eds) Advances in design, simulation and manufacturing III. DSMIE 2020. Lecture notes in mechanical engineering. Springer, Cham, pp 118–127. https://doi.org/10.1007/978-3-030-50794-7_12

Ivanov V, Liaposhchenko O, Denysenko Y, Pavlenko I (2021) Ensuring economic efficiency of flexible fixtures in multiproduct manufacturing. Eng Manag Product Serv 13(1):53–62. https://doi.org/10.2478/emj-2021-0004

Karpus VE, Ivanov VO, Kotliar OV, Minenko DO, Ivanova MS (2012) Intensification of manufacturing processes. Sumy, Sumy State University, p 436. https://doi.org/10.13140/2.1.3953.7604

Karpus VE, Ivanov VA (2012a) Choice of the optimal configuration of modular reusable fixtures. Russ Eng Res 32:213–219. https://doi.org/10.3103/S1068798X12030124

Karpus VE, Ivanov VA (2012b) Locating accuracy of shafts in V-blocks. Russ Eng Res 32:144–150. https://doi.org/10.3103/S1068798X1202013X

Karpus' VE, Ivanov VA (2008) Universal-composite adjustable machine-tool attachments. Russian Eng Res 28:1077–1083. https://doi.org/10.3103/S1068798X08110105

Krol O, Sokolov V, Logunov O, Shuliak M (2024) Development of a modular slotting head with a modified rack-and-pinion transmission. Int J Mech Appl Mech 16:146–155. https://doi.org/10.17683/ijomam/issue16.17

Kusyi Y, Stupnytskyy V (2020) Optimization of the technological process based on analysis of technological damageability of castings. In: Ivanov V, Trojanowska J, Pavlenko I, Zajac J, Peraković D (eds) Advances in design, simulation and manufacturing III. DSMIE 2020. Lecture notes in mechanical engineering, Springer, Cham, pp 276–284. https://doi.org/10.1007/978-3-030-50794-7_27

Kusyi Y, Onysko O, Kuk A, Solohub B, Kopei V (2022) Development of the technique for designing rational routes of the functional surfaces processing of products. In: Karabegović I, Kovačević A, Mandžuka S (eds) New technologies, development and application V. NT 2022. Lecture notes in networks and systems 472. Springer, Cham, pp 135–143. https://doi.org/10.1007/978-3-031-05230-9_16

Kusyi Y, Kusen N, Slipchuk A, Schuliar I, Pituley L (2024) Design of the technological route for sustainable machining of functional surfaces for automotive engineering parts. In: Ivanov V, Trojanowska J, Pavlenko I, Rauch E, Piteľ J (eds) Advances in design, simulation and manufacturing VII. DSMIE 2024. Lecture notes in mechanical engineering, Springer, Cham, pp 263–275. https://doi.org/10.1007/978-3-031-61797-3_22

Lieposhkina N, Denysenko Y, Trojanowska J, Korotun M, Pop G (2024) Statistical methods for decision-making regarding technological systems. In: Burduk A, Batako ADL, Machado J, Wyczółkowski R, Dostatni E, Rojek I (eds) Intelligent systems in production engineering and maintenance III. ISPEM 2023. Lecture notes in mechanical engineering, Springer, Cham, pp 293–305. https://doi.org/10.1007/978-3-031-44282-7_23

Petrakov YV, Ohrimenko OA, Sikailo MO, Myhovych AV (2023) Cutting forces simulation for end milling. J Eng Sci (Ukraine) 10(2):A27–A33. https://doi.org/10.21272/jes.2023.10(2).a4

Redko R, Zabolotnyi O, Redko O, Savchuk S, Kovalchuk V (2019) Improvement of manufacturing technology and recovery of clamping collets for lathe automats. In: Ivanov V et al (eds) Advances in design, simulation and manufacturing II. DSMIE 2019. Lecture notes in mechanical engineering, Springer, Cham, pp 290–301. https://doi.org/10.1007/978-3-030-22365-6_29

Silva H, Santos AS, Varela LR, Trojanowska J, Berladir K (2023) Design of experiments: an overview and future paths. In: Burduk A, Batako A, Machado J, Wyczółkowski R, Antosz K, Gola A (eds) Advances in production. ISPEM 2023. Lecture notes in networks and systems, vol 790. Springer, Cham, pp 328–341. https://doi.org/10.1007/978-3-031-45021-1_25

Trojanowska J, Kaščak J, Husár J, Knapčíková L (2022) Possibilities of increasing production efficiency by implementing elements of augmented reality. Bull Polish Acad Sci Tech Sci 70(6):e143831. https://doi.org/10.24425/bpasts.2022.143831

Yakovenko I, Permyakov A, Ivanova M, Basova Y, Shepeliev D (2021) Lifecycle management of modular machine tools. In: Tonkonogyi V, Ivanov V, Trojanowska J, Oborskyi G, Pavlenko I (eds) Advanced manufacturing processes III. InterPartner 2021. Lecture notes in mechanical engineering, Springer, Cham, pp 127–137. https://doi.org/10.1007/978-3-030-91327-4_13

Zakharkin O (2004) Technological fundamentals of manufacturing engineering. Sumy State University, Sumy (2004). ISBN 966-657-019-X [in Ukrainian]

Zaleta OM, Povstyanoy OY, Ribeiro LF, Redko RG, Bozhko TY, Chetverzhuk TI (2023) Automation of optimization synthesis for modular technological equipment. J Eng Sci 10(1):A6–A14. https://doi.org/10.21272/jes.2023.10(1).a2

Chapter 6
Machining Methods

6.1 Turning

Turning forms surfaces of revolution on machines of a turning group: turning and screw-cutting machines, turning rotary-table machines, CNC turning machines, automatic turret lathes, semi-automatic machines, and automatic machines with different numbers of spindles, multi-tool lathes, and hydraulic copying machines. All these machines have their limits of economic feasibility for different types of production (Kusyi and Stupnytskyy 2020; Korkmaz et al. 2024). The cutting tools for turning are single-blade tools. Examples of turning tools and surfaces to be turned are shown in Fig. 6.1.

Irrespective of the machine type, the principal turning motion is provided by workpiece rotation at the cutting speed V (m/min):

$$V = \frac{\pi \cdot D \cdot n}{1000}, \tag{6.1}$$

where

D a machined surface diameter, mm;
n frequency of workpiece rotation, rev/min.

The cutting speed during turning depends on the machining stage, workpiece material, and a tool's cutting part, averaging 50–250 m/min. The feed rate S during turning is provided by moving a tool along the workpiece axis, perpendicular to it, at an angle, or along a complex path when machining complex (shaped) surfaces. It is measured in millimeters per revolution (mm/rev).

Technologically, blade machining of center hole surfaces can be performed on turning machines in combination with the machining of outer surfaces and ends, increasing the accuracy of their relative positioning (Gasanov et al. 2019).

V. Ivanov et al., *Fundamentals of Manufacturing Engineering Using Digital Visualization*, Springer Tracts in Mechanical Engineering,
https://doi.org/10.1007/978-3-031-74360-3_6

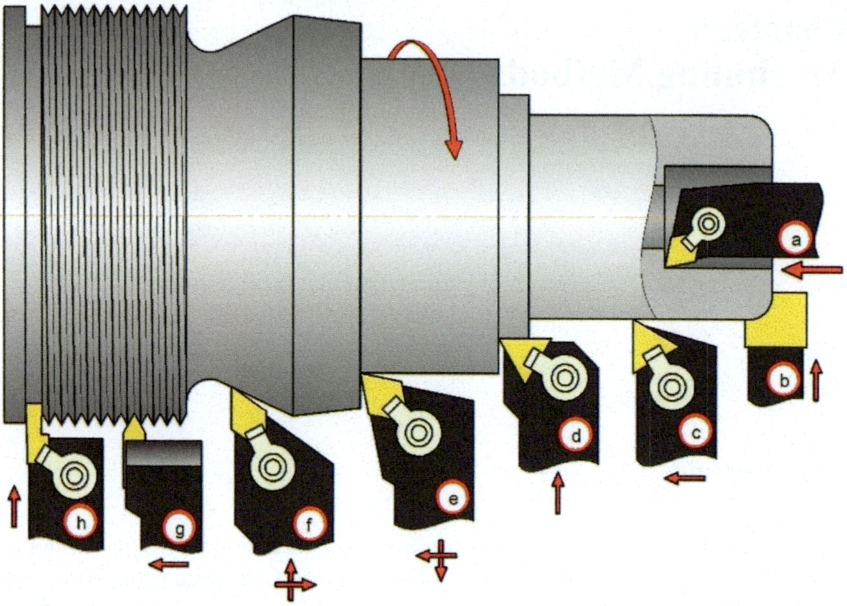

Fig. 6.1 Examples of tools and surfaces machined by turning: **a**—longitudinal boring of a cylindrical hole with a boring cutter; **b**—profiling with a form cutter; **c**—longitudinal turning of a cylindrical surface with a thrust-and-facing cutter; **d**—end trimming with a thrust cutter; **e**—longitudinal turning of a cylindrical surface with end trimming with a passing cutter; **f**—profiling and (or) turning of a conical surface with a contour cutter; **g**—cutting a thread with a threading cutter; **h**—grooving or forming a flute, or cutting with a grooving cutter or cutoff cutter

Turning ensures surface accuracy of 12–14th class of quality with a surface roughness of 25–6.3 μm according to the R_a criterion in case of rough machining to 6–7th class of quality during fine turning with a roughness of 0.2–1.6 μm according to the R_a criterion.

Hole boring is performed mainly on machines of turning, boring, and milling groups. Boring is performed with various designs of tools mounted either directly in a machine tool holder or in special mandrels (Orgiyan et al. 2020). Workpiece rotation provides the cutting speed, and tool movement provides the feed rate when boring a hole in a turning machine. For boring group machines, the tool rotates while a spindle provides feed with the tool and a table with a workpiece (Balaniuk et al. 2023). Multi-blade boring heads and blocks increase labor productivity when boring holes in medium-batch and large-batch manufacturing (Badovskyi et al. 2024).

Similar to turning, boring provides surface the accuracy of 12–14th grade class of quality with a surface roughness of 25–6.3 μm according to the R_a criterion in the case of rough machining to the 6–7th grade class of quality in fine turning with the surface roughness of 0.2–1.6 μm according to R_a.

6.2 Drilling, Counterboring, Reaming

Various off-center holes in body parts and radial holes in parts of bodies of revolution are typically machined on drilling and milling machines. Hole forming has certain technological peculiarities, including difficulty in chip extraction, significant heating of a workpiece and a tool, impossibility of monitoring a cutting zone, etc.

Blade machining of holes can be performed with end-cutting (to-size) tools, such as drills, countersinks, and reamers, and off-size tools, such as cutters. Some schemes for machining using to-size tools are shown in Fig. 6.2. They can be implemented on machines with vertical or horizontal spindle axes.

Drilling is the only method of blade machining for creating a hole in a solid material (Fig. 6.2a). Drills are cutting tools used and can be designed for different purposes. Twist drills are used to make holes in solid material (Fig. 6.2a) and to increase the size of an already drilled hole (reaming, Fig. 6.2b).

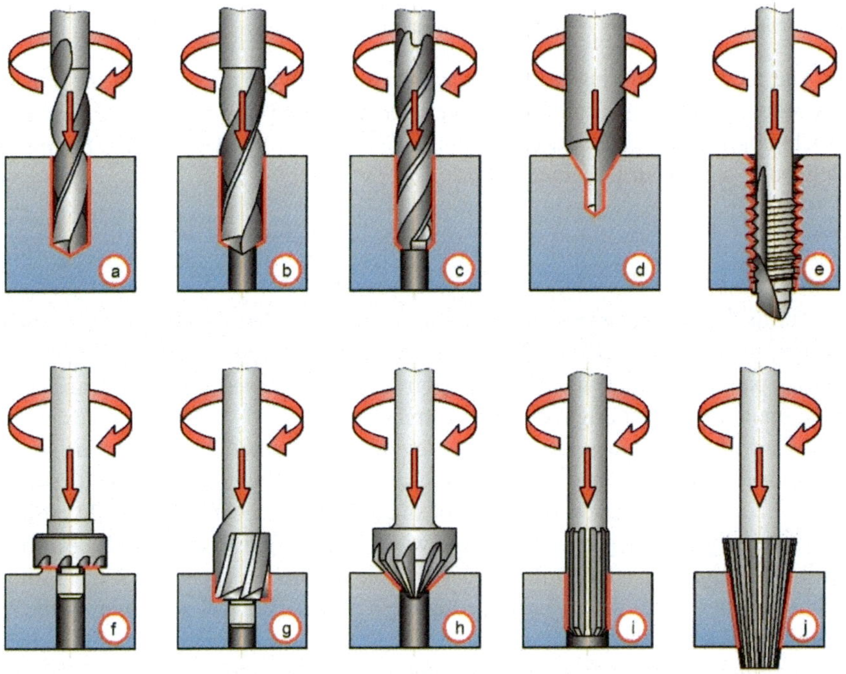

Fig. 6.2 Examples of hole surfaces machined with a to-size axial tool: **a**—drilling a cylindrical hole in a solid material with a twist drill; **b**—reaming a cylindrical hole with a twist drill; **c**—counterboring a cylindrical hole with a countersink; **d**—centering with a combined (centering) drill; **e**—cutting a thread with a tap; **f**—spotfacing a flat surface of a protrusion with a counterbore; **g**—spotfacing a cylindrical hole with a flat face with a spotfacing cutter; **h**—counterboring a chamfer with a countersink; **i**—reaming a cylindrical hole with a cylindrical reamer; **j**—reaming a conical hole with a conical reamer

Combined axial tools are designed to create multiple surfaces in a single stroke. They can combine other tools in one design, such as a countersink drill, etc.

Trepanning drills are used for machining holes with a large diameter (usually more than 60–70 mm). After the hole is made, a rod remains for further manufacturing.

The drills are made of high-speed types of steel and are also equipped with carbide inserts.

The cutting speed at drilling depends on a machining step, workpiece material, and a cutting part of the tool; on average, it is $V = (5\text{–}80)$ m/min. Depending on the machine, it is achieved by rotating a drill (drilling and boring machines) or a workpiece (lathes). The cutting speed can be determined by the formula

$$V = \frac{\pi \cdot D_{dr} \cdot n}{1000}, \tag{6.2}$$

where

D_{dr} a drill diameter, mm;
n rotation frequency of a drill or a workpiece, rev/min.

The feed rate S during drilling is achieved by moving a tool along the axis of a workpiece and is measured in mm/rev per revolution of a rotating element of a machining system (tool or workpiece). When drilling, the depth of cut t (mm) is defined as half the drill diameter D_{dr}:

$$t = \frac{D_{dr}}{2}. \tag{6.3}$$

The drilling diameter in a solid material is usually 27–30 mm. Drills of a particular design, like cannon, ejectors, rings, etc., are used for deep holes. A coolant cools a cutting part of the drill and removes chips from a cutting zone (Kolesnyk et al. 2022). It is supplied to the cutting zone under a pressure of about 5–8 MPa. In some designs of deep drilling machines, the drill can provide the main motion at the cutting speed. Deep drilling is performed at cutting speeds of up to 120 m/min and feed rates of up to 0.2 mm/rev. When drilling holes of small diameter, technology should provide a preliminary centering of a future hole with a centering drill (Fig. 6.2d) or a short drill of a larger diameter. Drilling ensures hole machining accuracy of the 11th to 13th class of quality with a surface roughness of 3.2–12.5 μm according to the R_a criterion (Yaşar et al. 2021).

Reaming (Fig. 6.2b) increases the hole size with a more significant drill bit. The cutting modes—cutting speed, feed rate, hole machining accuracy, and surface roughness—correspond to conventional drilling.

Counterboring (Fig. 6.2c) is used to finish pre-formed holes (drilled or produced in a workpiece by casting, stamping, etc.) to increase the size and improve accuracy.

A countersink is a multi-blade tool that is used as a cutting tool. Countersinks come in a variety of designs, from solid to tipped. The cutting part can be made of high-speed steel or with hard-alloy inserts.

The cutting speed during countersinking depends on the workpiece material and the tool's cutting part with the average V speed = 30–80 m/min at the feed rate of S = 0.5–1.5 mm/rev.

Tapered holes can be counterbored for reaming with a taper reamer. Counterboring ensures hole machining accuracy of the 8–10th class of quality with a surface roughness of 1.25–3.2 μm according to the R_a criterion. Counterboring is used to make tapered cavities (chamfers) in holes (Fig. 6.2h) for tapered countersunk screw heads, when cutting a thread, to blunt sharp edges, etc. Counterboring can be done with specific tools, such as countersinks, or with large-diameter drills that are sharpened at the desired angle. Combined tools like drill countersinks are widely used in large-batch and mass manufacturing.

Spotfacing involves machining an external hole face (Fig. 6.2f) or step hole face (Fig. 6.2g) to fit bolt heads, etc., so that they are perpendicular to the hole axis. Spotfacing is used to machine hole faces on drilling and boring machines. Spotfacing is a counterboring type performed with a spotfacer—a face-cutting tool. A spotfacer has a guiding element designed to orientate its axis in a pre-made hole. Spotfacers are widely used in large-batch and mass manufacturing.

Reamers are mounted in special "floating" chucks when machining fine holes on machine tools to compensate for the misalignment of the machine spindle axis with the hole axis to be machined. The reaming cutting speed is $V = 2–15$ m/min at the feed rate $S = 0.5–3.5$ mm/rev. Finishing reaming ensures hole machining accuracy of the 6–7th class of quality with a surface roughness of 0.63–1.25 μm according to the R_a criterion.

Internal thread with taps (Fig. 6.2e) demands more labor due to the need to reverse the tool (or workpiece) after threading to return the tool to its original position. Tapping is used for cutting threads of relatively small diameter (up to 52 mm) and low accuracy (7–8th degree) with a surface roughness of 1.6–3.2 μm according to the R_a criterion. Tapping can be performed manually and on machine tools for turning (Onysko et al. 2023), drilling, boring groups, bolt-cutting machines, etc. Taps are also used to cut tapered threads (Onysko et al. 2024).

Manual threading usually involves a set of two or three taps. Machine taps, as a rule, form a thread in one working stroke, i.e., with one tool, not a set. In large-batch and mass manufacturing, tap designs allow the tap to be withdrawn from the hole without reversal.

Machine taps are mounted in special floating holders that compensate for deviations from hole-tap alignment. To prevent the tap from breaking when cutting in blind holes, the tap holders are equipped with an over-torque safety mechanism. When the tap rests on the bottom of the hole, this mechanism decouples the tap rotation from machine spindle rotation.

Cutting speeds for tapping depend on the material of a workpiece, the size of a cut, etc., and range from 0.5 to 15 m/min. Taps require mandatory surface lubrication with oils, and special lubricant mixtures are used when machining hard-to-machine materials. Taps with a ground cut profile increase cutting accuracy by up to 5–6 degrees with a surface roughness of 1.25–1.6 μm according to the R_a criterion.

6.3 Milling

Milling is performed using multi-blade tools—milling cutters, the design of which allows the machining of planes (both in horizontal and vertical positions at an angle to a machine table plane), cylindrical surfaces, groove surfaces, and shaped elements (Figs. 6.3, 6.4 and 6.5).

Milling is carried out on vertical and horizontal milling, horizontal boring, mill-drill-bore machines, carousel and drum milling machines, and CNC machining centers (Permyakov et al. 2019).

The main cutting movement with the cutting speed of V (m/min) is provided by the cutter's rotational movement:

$$V = \frac{\pi \cdot D_{mc} \cdot n}{1000}, \quad (6.4)$$

where

D_{mc} a milling cutter diameter, mm;
n rotation frequency of a milling cutter, rev/min.

The feed S, usually a relative translational motion of a workpiece and a cutter during milling, can be of the following types: feed per tooth S_z, mm/tooth; feed rate per revolution $S_{rev} = S_z \cdot$ z, mm/rev (z is several cutter teeth); minute feed rate $S_{min} = S_{rev} \cdot n$, mm/min.

Depending on the design of a cutting part in a milling cutter, its material, and the machining conditions, milling provides an accuracy of the 8–14th class of quality with a surface roughness of 0.8–12.5 μm according to the R_a criterion (Duplak et al. 2021).

Milling is a highly productive machining method used in almost all types of production. Milling of external cylindrical surfaces is a productive method of machining. It is used for machining, for example, the journals of stepped- and crank-type shafts, brake plates, etc., performed with disk, end mill, or face milling cutters on machines of the milling group. Feed motion is usually provided using additional devices by rotating a workpiece around its axis at a speed of 10–20 m/min (rotary feed). In contrast, the cutting speed V (m/min) is provided by the rotational speed of the cutter (main cutting motion). When a disk cutter is used, its width is equal to the width of the surface to be machined. The method ensures machining accuracy of the 9–10th class of quality with a surface roughness of 5–8 microns according to the R_a criterion (Duplak et al. 2023).

Milling is one of the most common methods of machining grooves, which can be shaped in different ways depending on the functional purpose: key slots, T-slots, dovetail slots, etc. Slotting is performed on machines of the milling and milling-boring groups. The groove is further processed by grinding, scraping, etc., to achieve high dimensional accuracy after milling.

When milling slots, the cutter's rotational movement is provided by the cutting speed V (m/min). Feed rate—the rate of relative movement of a workpiece and a tool

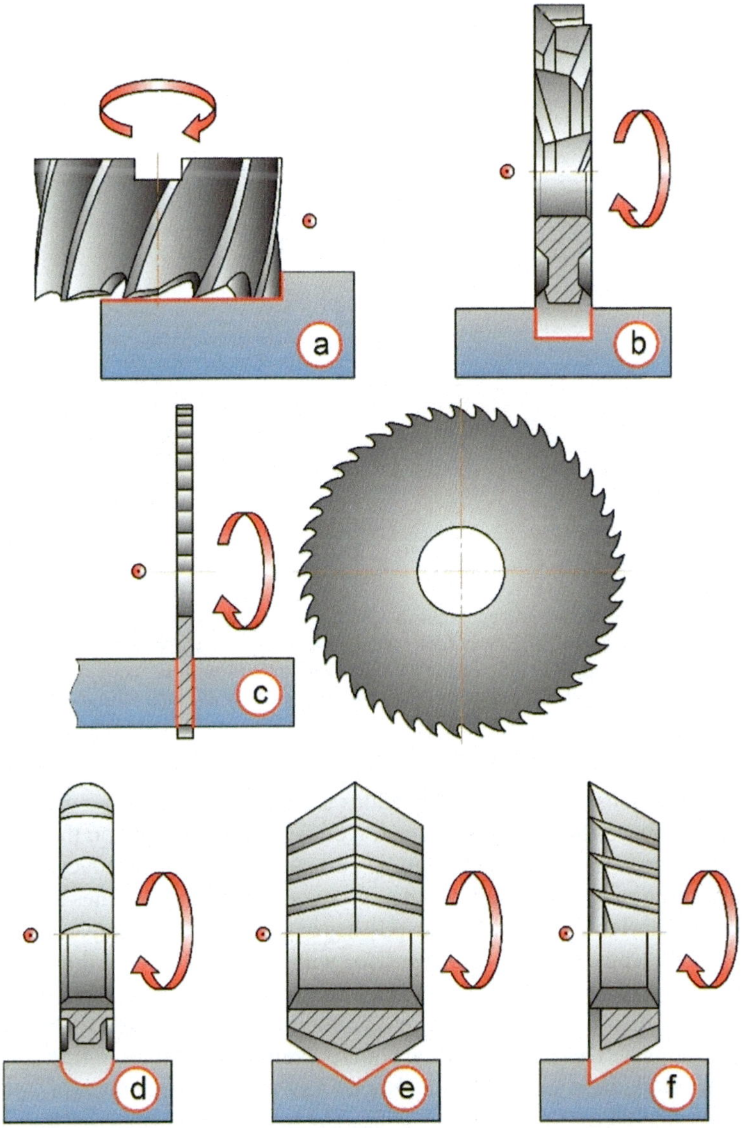

Fig. 6.3 Surface machining with high-speed cutters and milling cutters with brazed carbide inserts: **a**—milling a plane (shoulder) with a face milling cutter; **b**—milling a U-shaped groove with a three-sided disk cutter; **c**—cutting off with a slitting disk cutter; **d–f**—milling of shaped grooves with cutters of semi-circular, prismatic, and angular profiles, respectively

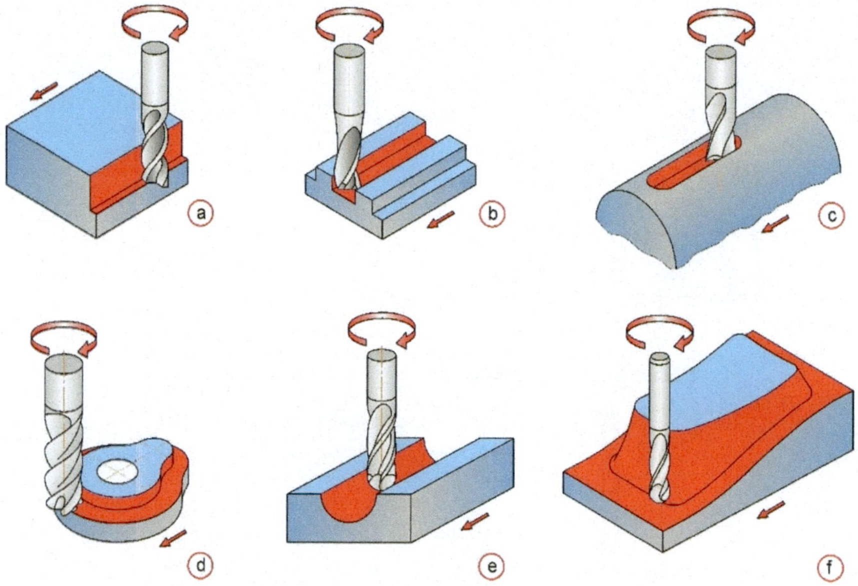

Fig. 6.4 Surface machining with end mills: **a**—milling a shoulder with an end mill; **b**—milling an open U-shaped groove with a keyway (cotter) cutter; **c**—milling a closed key slot with a keyway (cotter) cutter; **d**—contour milling of the shoulder with an end mill; **e**—copy milling or pocket milling with a radius or spherical cutter; **f**—copy milling or cavity milling with a radius or spherical cutter

S (mm/min) in most machining applications is achieved by moving a machine table with a workpiece. The cutting speed and feed rate values depend on the workpiece material, the material of the tool's cutting part, and machining accuracy.

When machining T-slots and dovetail grooves, machining schemes are used that involve pre-machining a rectangular (U-shaped) groove, for example, with a disk cutter, and then machining a shaped surface with a special T-slot or dovetail cutter.

Milling is also used to process threaded surfaces (Neshta et al. 2018). Typically, threads are milled on specialized thread milling machines as well as on universal milling machines. This method is used both for preliminary thread cutting and subsequent machining on a lathe (for threads with a sufficiently large pitch of 2–12 mm) and as the method for final thread shaping. The machine mills external and internal cuts of triangular, rectangular, trapezoidal, and other profiles.

There are two significant methods of thread milling: disk milling and group milling. A disk milling cutter is set at an angle α equal to the angle of a thread helix. The cutting speed provides the cutter's rotational movement. In addition, a workpiece rotates at the rotary feed rate, and the workpiece's and the cutter's relative axial movement (per workpiece revolution) is achieved by the value of the cutting pitch. Disk milling can be used to machine long cuts.

Group milling is used for milling external and internal single- and multiple-start threads. Milling is performed on special cutter milling machines. This machining

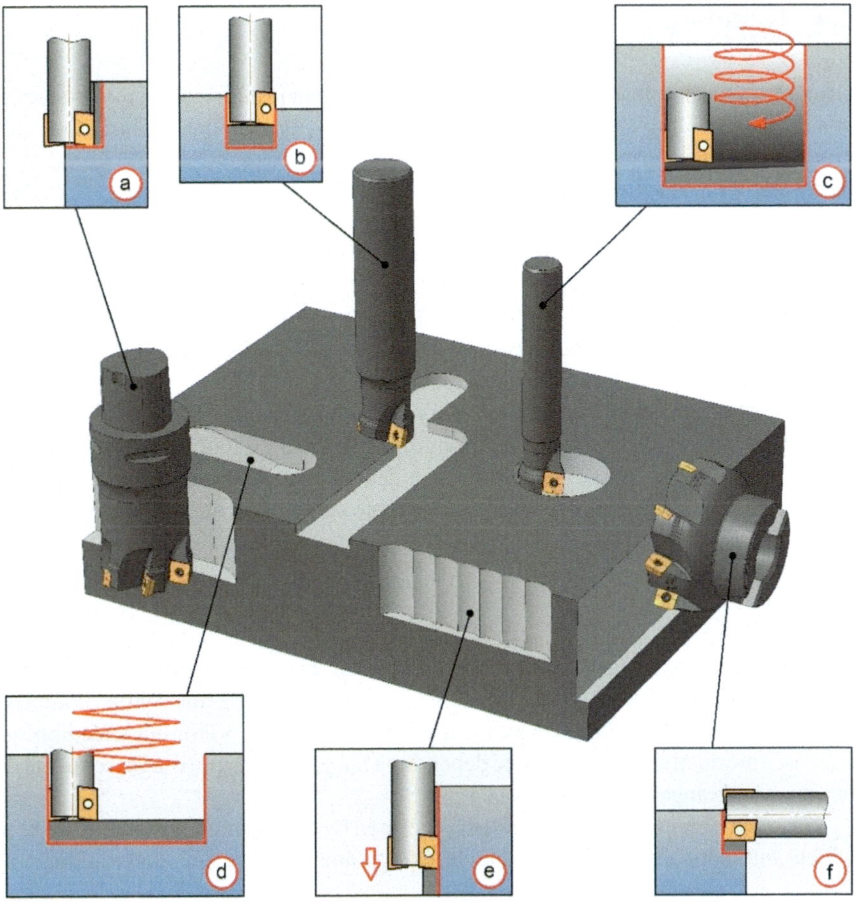

Fig. 6.5 Surface treatment with milling cutters with replaceable cutting inserts: **a**—milling a shoulder; **b**—milling a groove; **c**—milling a cavity (pocket) with a helical interpolation path; **d**—milling a cavity (pocket) with an inclined penetration; **e**—penetrating milling; **f**—face milling

method requires the cutter width to be larger than the thread length. The workpiece is machined per $1\frac{1}{4}$ of the workpiece revolution (1/4 revolution of the workpiece is required for a cutter to plunge with the radial feed to the thread depth).

This method also allows tapered and multi-start threads to be made. Group milling is a very productive machining method used in large-batch and mass manufacturing. The thread accuracy is the 5–7th degree with a roughness of 0.63–3.2 μm according to the R_a criterion.

6.4 Grinding

Grinding is a technological method of surface treatment using a cutting tool—a wheel consisting of cutting elements in the form of abrasive or diamond grains. Planes are ground on surface grinding machines with a rectangular or roundtable. Grinding is performed both on the periphery and on the wheel face. Typical surface grinding patterns are shown in Figs. 6.6 and 6.7.

In grinding, the cutting speed V is provided by a grinding wheel and is measured in m/s, unlike in blade machining. On circular table machines, the table rotates continuously, and workpieces are mounted and removed after machining as they pass through the rotational plane in a certain section. These machines are very productive and are used for large volumes of production.

When grinding with the wheel periphery on rectangular table machines (Fig. 6.6a), the table moves reciprocating with the longitudinal feed $S_l = 20–60$ m/min. The cross-feed S_{cr} (mm) must be provided if the machined width exceeds the wheel width. The cross-feed rate is calculated for one double table stroke. It is measured as a fraction of the disk width B:

$$S_{cr} = (0.2 - 0.7) \cdot B. \tag{6.5}$$

Once the entire surface width has been ground, the plunge feed S_{in} is applied to provide the depth of penetration. For disk grinding (Fig. 6.6b), disks with a diameter larger than the width of a workpiece are used. This eliminates preliminary feeding and increases productivity. The feed rate depends on the grinding type (rough, semi-finish, or other) and ranges from 0.015–0.04 mm.

The table's reciprocating movement is ensured by the machine's hydraulic system, which automatically reverses the table using cams mounted on it, switching the

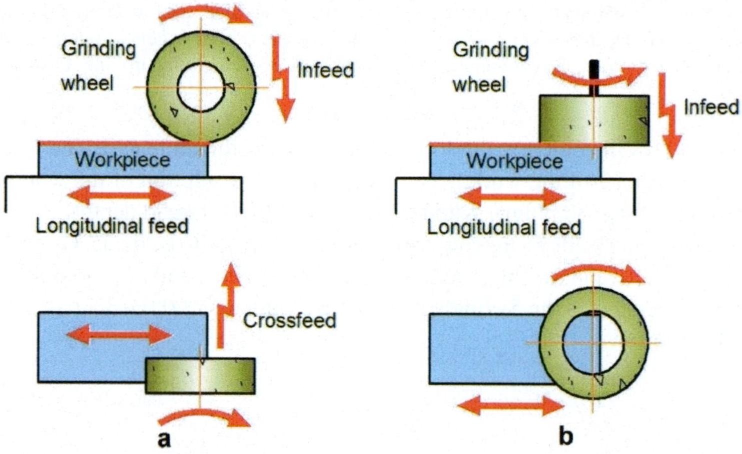

Fig. 6.6 Grinding flat surfaces on a rectangular table: **a**—peripheral grinding; **b**—side grinding

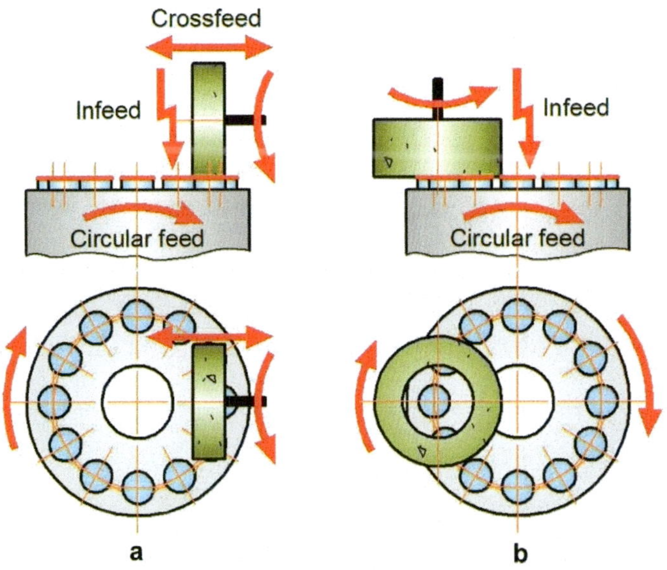

Fig. 6.7 Grinding flat surfaces on a round table: **a**—peripheral grinding; **b**—side grinding

directional valve. On circular table machines (Fig. 6.7), the circular feed rate is determined by the rotational speed of the table itself. It is set in the $S_c = 20\text{–}60$ m/min range depending on machining conditions. The cross-feed rate on these machines ranges from:

$$S_{cr} = (0.2 - 0.6) \cdot B. \tag{6.6}$$

Surface grinding provides the 6–7th dimensional accuracy grades with a surface roughness of 0.63–0.16 μm according to the R_a criterion. Surface grinding machines are equipped with magnetic plates (permanent magnets or electromagnets), allowing for easy clamping of workpieces. If the material of a workpiece is not magnetic, various devices are used to fix it. Grinding generates a considerable amount of heat, which is why it is imperative to use coolant lubricants (Stepanov et al. 2021).

When grinding external cylindrical surfaces, there are two basic grinding schemes: longitudinal grinding (Fig. 6.8a) and cross-grinding (plunge grinding) (Fig. 6.8b).

Grinding is performed on cylindrical grinding machines (Stepanov et al. 2021). The cutting speed $V = 35\text{–}60$ m/s is achieved by rotating a grinding wheel. A workpiece is usually mounted in solid centers and rotates at the speed of 10–20 m/min. When grinding with longitudinal feed, a table moves relative to the wheel with the feed rate S_l, measured as a fraction of an abrasive disk width per revolution of a workpiece:

$$S_l = k \times B_w, \tag{6.7}$$

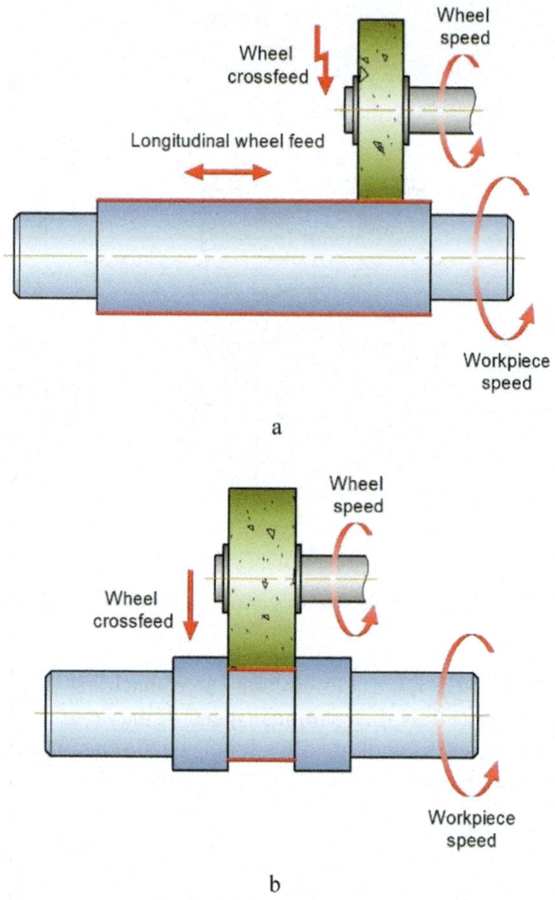

Fig. 6.8 Grinding external cylindrical surfaces: **a**—longitudinal grinding; **b**—cross (infeed)-grinding

where

k a fractions number of a circle width ($k = 0.2–0.7$);
B_w a circle width, mm.

After the wheel has passed along the entire length of the surface L_p, the cross-feed S_{cr} is applied, which is the depth of cut. Then, the cycle of movements is repeated. Depending on the machining stage (rough, semi-finish, etc.), the cross-feed rate S_{cr} ranges from 0.005 to 0.06 mm. At the end of machining, when the entire allowance has been removed, two to three more passes are made along the surface without cross-feed until sparking stops (sparking out). This technique improves accuracy by removing shape errors caused by elastic deformations of the machining system.

Cross-feed grinding is performed with abrasive disks that are wider than the length of the surface to be machined. Machining is performed only with the cross-feed rate of a wheel Scr, measured in mm/revolution of a workpiece. Its value is in the range of 0.003–0.02 mm/rev.

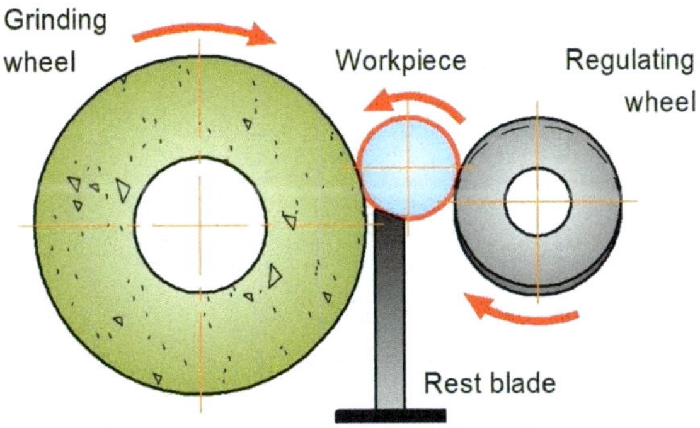

Fig. 6.9 External centerless grinding

Plunge grinding can be performed with several disks simultaneously, allowing for grinding shaped surfaces (if a disk with the appropriate shaped profile is available). This type of grinding is widely used in large-batch and mass manufacturing (Syzyi et al. 2020).

The accuracy of finish grinding of surfaces reaches the 6–7th class of quality with a surface roughness of 0.2 μm according to the R_a criterion.

In addition to the methods described above, centerless grinding (Fig. 6.9) is used for parts such as pins and smooth and stepped rollers, and it can also be performed by passing and plunge grinding.

The following basic grinding schemes are used when grinding holes (Figs. 6.10 and 6.11). Holes are ground on internal grinding machines. The cutting speed of $V = 35$–60 m/s is achieved by rotating a grinding wheel. When grinding holes of small diameter (less than 10 mm), it is difficult to achieve such cutting speeds so that they can be around 10 m/s. When grinding with longitudinal feed (Fig. 6.10a), a workpiece is usually mounted in a chuck and rotates at 10–20 m/min. The kinematics of the process are similar to external grinding.

A wheeled grinding head moves relative to a workpiece at the feed rate S_l, which is determined as a fraction of the width of the grinding wheel per revolution of the workpiece. After the wheel has passed along the entire length of the surface L_p, the feed rate S_{cr} is applied, which is the depth of cut. Then, the cycle of movements is repeated. Depending on the machining stage (roughing, semi-finish, etc.), the cross-feed rate S_{cr} ranges from 0.003 to 0.015 mm. At the end of machining, the tool is sparked out.

Plunge grinding with the cross-feed is performed with grinding wheels that are wider than the length of the surface to be processed. Machining is performed only with the cross-feed of the wheel S_{cr}, measured in mm/revolution of a workpiece. Its value is within the range of 0.003–0.02 mm/rev. Plunge grinding allows for the grinding of shaped surfaces (if a wheel with the appropriate shaped profile is available). This type of grinding is widely used in large-batch and mass manufacturing. A planetary grinding scheme (Fig. 6.10b) is used for large workpieces that cannot be rotated. In this scheme, the grinding wheel is given additional rotational motion (planetary motion) and its axis to produce a desired hole size.

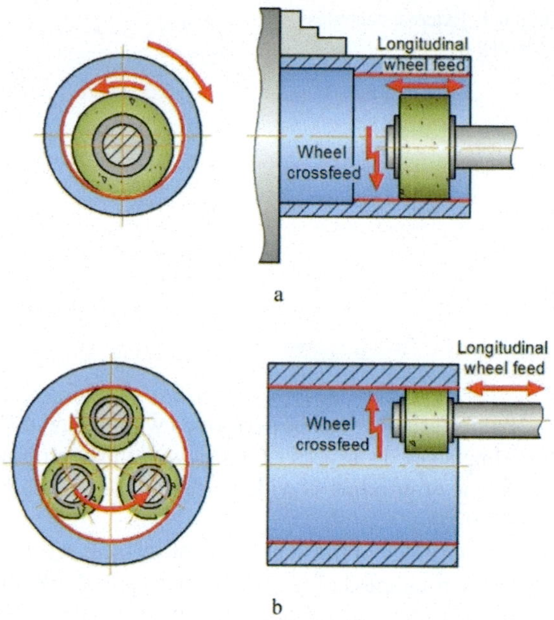

Fig. 6.10 Internal grinding: **a**—longitudinal grinding; **b**—planetary grinding

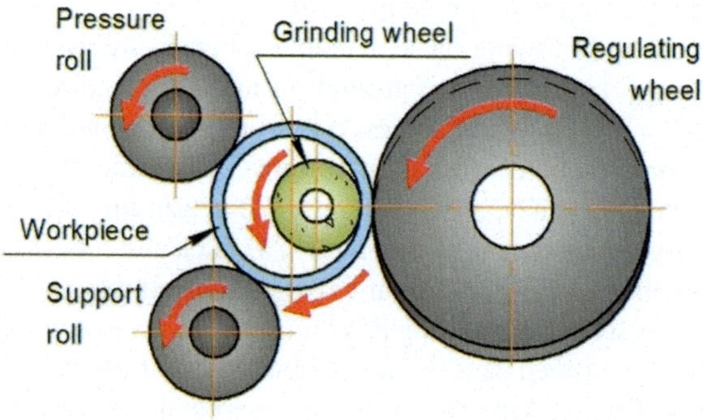

Fig. 6.11 Internal centerless grinding

Centerless grinding (Fig. 6.11) involves the rotation of a loose workpiece. Since the outer surface is also a process datum, prior precision grinding is required. This method ensures the highest accuracy of the bushings' bore-to-external surface alignment (up to 0.003 mm). The finish grinding accuracy reaches the 6–7th grade of quality with a surface roughness of 0.32–1.6 μm according to the R_a criterion.

6.5 Gear Machining

Modern machines use gears to transmit torque between parallel shafts and shafts positioned at a certain angle. The most common types of gears are cylindrical, bevel, and worm.

There are 12 degrees of accuracy for spur gear wheels. Each grade is specified regarding kinetic accuracy, smoothness, tooth contact, and six combinations based on the guaranteed backlash. Gear wheels of 6–8 degrees of accuracy are most common in general engineering and machine building.

Gear teeth are machined on milling machines (for single and small-batch manufacturing) and special gear-cutting machines.

Both blade machining (milling, slotting, shaving) and abrasive machining (grinding, grinding in) are used for machining.

In gear machining, two fundamentally different methods are used—copying and rolling. These methods are characterized by different precision and machining performance. The first method involves machining with a tool with a gear cavity profile and a subsequent rotation of a workpiece at a certain angle, depending on the number of teeth. This method is used in repair production without special equipment, in manufacturing large gears, and for pre-cutting teeth for further precision machining. The second method simulates the operation of a gear pair. In this case, one element in the pair acts as a cutting tool, and the other is a workpiece.

Gear teeth milling is performed using either the copying or the rolling method. The copying method can be implemented with involute disks and pencil cutters (Fig. 6.12).

It enables the production of spur, chevron, and bevel gears. Modular disk cutters are available in 8, 15, or 26 pieces. An 8-piece set is used for wheels with a modularity

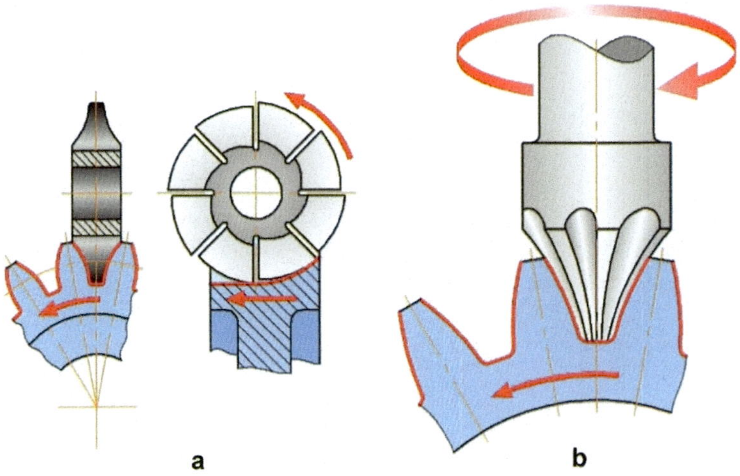

Fig. 6.12 Gear milling by involute gear cutters: **a**—disk mill; **b**—pencil mill

of up to 9 mm. Each cutter can deliver the required precision when machining wheels with a specific number of teeth. For example, cutter No. 1 in the 8-piece set is designed for machining wheels with a tooth count of 35 to 54. A certain cutting speed is ensured by rotating a cutter and lies in the range of 25–40 m/min. Number of working strokes depends on the wheel module (with $m < 6$ mm—in one working stroke, with $m =$ 6–12 mm—in two, and with $m > 12$—in three). Pencil mill cutters are best suited for wheels with a 10 to 50-mm module and for cutting chevron wheels without a groove. Cutting with modular cutters requires using a dividing head or table to set up a workpiece. This method ensures cutting wheels of the 9th-10th degree of accuracy with a surface roughness of 6.3–12.5 μm according to the R_a criterion.

The rolling method uses worm cutters on hobbing machines (Fig. 6.13a).

The cutting speed V (m/min) is provided by rotating a cutter. In addition, the machine movement ensures the workpiece rotation—S_c—and the cutter movement S_l along the workpiece axis. This is done by setting up the change-gear units of the machine (speed, feed, division, rolling, etc.).

Depending on the direction of the cutter's helical flutes (right or left) and the wheel type (spur, helical, right, or left), the cutter is set at a certain angle λ relative to the workpiece axis (Fig. 6.13b). The formula determines this angle

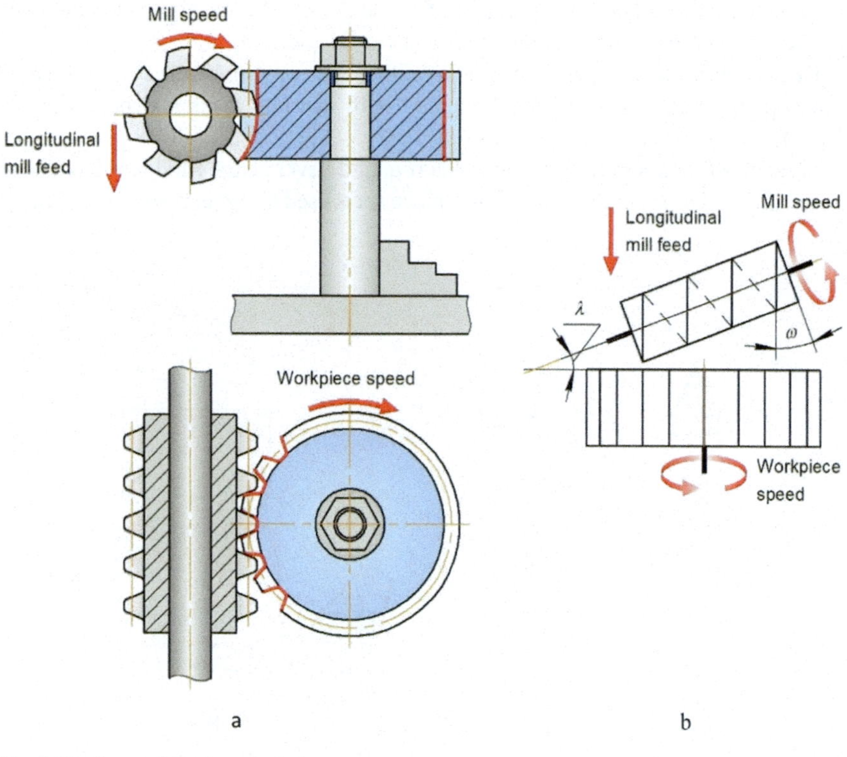

Fig. 6.13 Gear milling by gear hob cutter

$$\lambda = \beta \pm \omega, \tag{6.8}$$

where

β inclination angle of cutting wheel teeth, deg.;
ω lifting angle of hob threads, deg.

A plus sign is used if the angles are different, and a minus sign if they are the same.

Modular worm cutters are available in both high-speed steel and carbide inserts.

Gear milling allows achieving the 8th-10th degree of accuracy of wheels with a surface roughness of 3.2–6.3 μm according to the R_a criterion.

Gear shaping is a process of cutting gears with hobs using the generating method on gear shaping machines (Hrytsay and Stupnytskyy 2023). The chisel rotates around its axis and reciprocates along the axis of a workpiece. The workpiece, in turn, also rotates (Fig. 6.14).

This method enables machining teeth on both spur and helical gears for external and internal coupling and the small-toothed wheel rim in block gears. Gear shaping accuracy is higher than the gear milling and allows achieving the 6–7th grade with a surface roughness of 1.6–3.2 μm according to the R_a criterion (Hrytsay and Stupnytskyy 2021). Wheels with a module up to 2.5 mm are machined in one revolution of a workpiece, and wheels with a larger module can be machined in several revolutions. Wheel tooth shaping is used in medium-batch and large-batch manufacturing. Mass manufacturing applies gear shaping by copying special multi-cutter heads equipped with several cutters corresponding to the number of gear teeth.

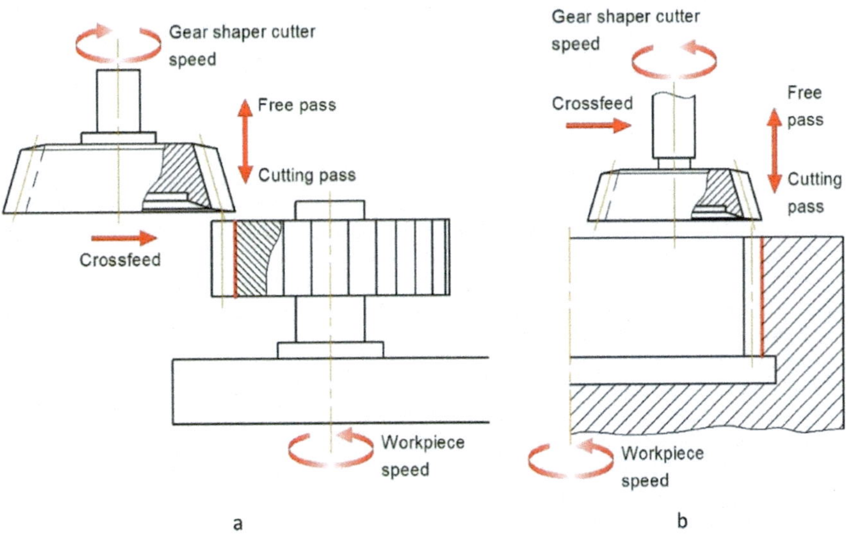

Fig. 6.14 Gear shaping by gear shaper cutter: **a**—external gear shaping; **b**—internal gear shaping

Gear shaving is a method of gear finishing without heat treatment. Gear shaving is performed on special shaving machines. A shaver is a cutting tool for this method and represents a gear with grooves cut on both sides of a tooth. These grooves create cutting edges that remove fine chips from the workpiece tooth surfaces during operation. This is achieved by positioning the workpiece and the shaver at a specific angle, allowing them to rotate and move along an axis.

The cutting speed V (100–120 m/min) is ensured by rotating the milling cutter. Their relative progressive movement—the feed rate S—is within 0.14–0.3 mm/revolution of the workpiece. Shaving ensures achievement of the 6–7th degree of wheel accuracy with a surface roughness of 0.63–1.25 μm according to the R_a criterion.

Grinding is performed on gear grinding machines using both copying and originating methods. This machining method provides the 5–6th degree of accuracy with the tooth surface roughness of 0.32–1.25 μm according to the R_a criterion. The copying method involves a profile wheel designed to grind a single cavity in several strokes along the tooth surface. The workpiece is then returned to the position for the next tooth. The workpiece is sometimes rotated by more than one tooth to increase accuracy. The generating method involves dish and ribbed grinding wheels. Machines are equipped with special devices for wheel adjustment.

Grinding-in is widely used in large-batch and mass manufacturing to produce heat-treated gears for critical gears. This method is based on grinding in a machined wheel interlocking with a lap. The requirements for the material and preparation for work with the lapping wheel (charging) are the same as those required for laps to other surfaces. On average, it takes 2–4 min to grind in one wheel. This process ensures the 5–6th degree of accuracy with the tooth surface roughness of 0.1–0.5 μm according to the R_a criterion.

References

Badovskyi O, Balaniuk A, Oborskyi G, Orgiyan A, Edl M (2024) An increase in the vibration resistance of finishing and boring machines when cutting ends using the plunge-in method. In: Ivanov V, Trojanowska J, Pavlenko I, Rauch E, Piteľ J (eds) Advances in design, simulation and manufacturing VII. DSMIE 2024. Lecture notes in mechanical engineering, Springer, Cham, pp 139–151. https://doi.org/10.1007/978-3-031-61797-3_12

Balaniuk A, Oborskyi G, Orgiyan A, Tonkonogyi V, Dašič P (2023) Dynamics of fine boring of intermittent surfaces. In: Moldovan L, Gligor A (eds) The 17th International conference interdisciplinarity in engineering. Inter-ENG 2023. Lecture notes in networks and systems, vol 926. Springer, Cham, pp 109–117. https://doi.org/10.1007/978-3-031-54664-8_11

Duplak J, Duplakova D, Zajac J (2023) Research on roughness and microhardness of C45 material using high-speed machining. Appl Sci 13:7851. https://doi.org/10.3390/app13137851

Duplak J, Duplakova D, Hatala M, Radchenko S, Sukic E (2021) Surveying the topography and examining the quality of the machined surface of selected hardened steels in the milling process. J Eng Res (Kuwait) 9(3):285–301. https://doi.org/10.36909/jer.v9i3B.9337

Gasanov M, Kotliar A, Basova Y, Ivanova M, Panamariova O (2019) Increasing of lathe equipment efficiency by application of gang-tool holder. In: Gapiński B, Szostak M, Ivanov V (eds)

Advances in manufacturing II. Manufacturing 2019. Lecture notes in mechanical engineering, Springer, Cham, pp 133–144. https://doi.org/10.1007/978-3-030-16943-5_12

Hrytsay I, Stupnytskyy V (2021) Prediction the durability of hobs based on contact and friction analysis on the faces for cutting teeth and edges during hobbing. In: Ivanov V, Trojanowska J, Pavlenko I, Zajac J, Peraković D (eds) Advances in design, simulation and manufacturing IV. DSMIE 2021. Lecture notes in mechanical engineering, Springer, Cham, pp 405–414. https://doi.org/10.1007/978-3-030-77719-7_40

Hrytsay I, Stupnytskyy V (2023) Hob wear prediction based on simulation of friction, heat fluxes, and cutting temperature. Archiv Mechan Eng 70(2): 271–286. https://doi.org/10.24425/ame.2023.145582

Kolesnyk V, Peterka J, Alekseev O, Neshta A, Xu J, Lysenko B, Sahul M, Martinovič J, Hrbal J (2022) Application of ANN for analysis of hole accuracy and drilling temperature when drilling CFRP/Ti alloy stacks. Materials 15:1940. https://doi.org/10.3390/ma15051940

Korkmaz ME, Gupta MK, Sarikaya M et al (2024) Analytical modeling methods in machining: a state of the art on application, recent challenges, and future trends. Arab J Sci Eng 49:10287–10326. https://doi.org/10.1007/s13369-024-09163-7

Kusyi Y, Stupnytskyy V (2020) Optimization of the technological process based on analysis of technological damageability of castings. In: Ivanov V, Trojanowska J, Pavlenko I, Zajac J, Peraković D (eds) Advances in design, simulation and manufacturing III. DSMIE 2020. Lecture notes in mechanical engineering, Springer, Cham, pp 276–284. https://doi.org/10.1007/978-3-030-50794-7_27

Neshta A, Kryvoruchko D, Hatala M, Ivanov V, Botko F, Radchenko S, Mital D (2018) Technological assurance of high-efficiency machining of internal rope threads on computer numerical control milling machines. ASME J Manufact Sci Eng 140(7):071012. https://doi.org/10.1115/1.4039062

Onysko O, Kopei V, Barz C, Kusyi Y, Baskutis S, Bembenek M, Dašić P, Panchuk V (2024) Analytical model of tapered thread made by turning from different machinability workpieces. Machines 12:313. https://doi.org/10.3390/machines12050313

Onysko O, Kopei V, Kusyi Y, Pituley L, Taras I (2023) Tool wear in the process of drill-string connector thread lathe machining. In: Karabegovic I, Kovačević A, Mandzuka S (eds) New technologies, development and application VI. NT 2023. Lecture notes in networks and systems, vol 687. Springer, Cham, pp 98–111. https://doi.org/10.1007/978-3-031-31066-9_11

Orgiyan A, Oborskyi G, Balaniuk A, Tonkonogyi V, Dasic P (2020) Development of calculation of statistical and dynamic errors upon fine boring with console boring bars. In: Tonkonogyi V et al (eds) Advanced manufacturing processes II. InterPartner 2020. Lecture notes in mechanical engineering, Springer, Cham, pp 588–597. https://doi.org/10.1007/978-3-030-68014-5_57

Permyakov A, Dobrotvorskiy S, Dobrovolska L, Basova Y, Ivanova M (2019) Computer modeling application for predicting of the passing of the high-speed milling machining hardened steel. In: Ivanov V et al (eds) Advances in design, simulation and manufacturing. DSMIE 2019. Lecture notes in mechanical engineering, Springer, Cham, pp 135–145. https://doi.org/10.1007/978-3-319-93587-4_15

Stepanov M, Ivanova M, Litovchenko P, Ivanova L, Havryliuk Y (2021) Study of heat transfer conditions in the cutting zone when grinding. In: Machado J, Soares F, Trojanowska J, Ottaviano E (eds) Innovations in mechanical engineering. ICIENG 2021. Lecture notes in mechanical engineering, Springer, Cham, pp 173–181. https://doi.org/10.1007/978-3-030-79165-0_17

Syzyi Y, Ushakov O, Slipchenko S, Basova Y, Ivanova M (2020) Simulation of the contact temperature in the cylindrical plunge grinding process. Diagnostyka 21(2):77–86. https://doi.org/10.29354/diag/122532

Yaşar N, Korkmaz ME, Gupta MK et al (2021) A novel method for improving drilling performance of CFRP/Ti6AL4V stacked materials. Int J Adv Manuf Technol 117:653–673. https://doi.org/10.1007/s00170-021-07758-0

SPRINGER NATURE

GPSR Compliance

The European Union's (EU) General Product Safety Regulation (GPSR) is a set of rules that requires consumer products to be safe and our obligations to ensure this.

If you have any concerns about our products, you can contact us on ProductSafety@springernature.com

In case Publisher is established outside the EU, the EU authorized representative is:

Springer Nature Customer Service Center GmbH
Europaplatz 3
69115 Heidelberg, Germany

The manufacturer's authorised representative in the EU is Springer
Nature Customer Service Centre GmbH, Europaplatz 3, 69115 Heidelberg,
Germany. If you have any concerns regarding our products, please
contact ProductSafety@springernature.com

Printed and bound by CPI Group (UK) Ltd, Croydon, CR0 4YY
30/09/2026
02225604-0003